Praise for *When the Uncertainty Pr*

"There are plenty of popular introductions to quantum mechanics, why read this one? The answer is obvious: this one goes up to eleven. The energy that Philip Moriarty puts into making connections between his two passions, quantum theory and heavy metal is a joy to behold. I defy anyone to find Fourier analysis dull when it is explained with reference to the span-dex strides of Stryper's drummer. You don't need to be a metalhead to like this book—but be warned that if you do like this book, you will probably find yourself more of a metalhead by the end than you were at the start, because the enthusiasm is infectious. You might even find you have a better grip of the notorious mind-warping concepts of quantum mechanics too."

—**PHILIP BALL**, author of *Beyond Weird: Why Everything You Thought You Knew about Quantum Physics Is Different*

"A magical mosh pit of Slayer and spandex trousers, sound waves, and strings—this is quantum physics as you've never seen or heard it before."

—**MATIN DURRANI**, editor of *Physics World* magazine and coauthor of *Furry Logic: the Physics of Animal Life*

"Both metal-heads and physicists have become caricatures in today's pop culture. In his wonderfully conversational writing, Moriarty smashes these stereotypes and subverts expectations by weaving the two worlds together. This book shows how unexpected ideas cut across the worlds of heavy metal and quantum physics. If you enjoy surprises, brutal band logos, or insane riffs, you'll love this book. Forgot pop-sci. This is metal-sci."

—**JESSE SILVERBERG, PhD**, physicist and Harvard research fellow

"I thought I'd already seen every possible analogy for the weird world of quantum physics, but Philip Moriarty's music-inspired take on it is fresh and engaging. You don't need to be a metal fan to enjoy this book. Moriarty's enthusiasm for both physics and metal shines through so much in his writing that I was tempted to break out the Megadeth myself while reading. If you've ever been intrigued by quantum mechanics but worried that you couldn't hack an entire book on the subject, try this one, and you won't be disappointed."

—**KELLY OAKES**, former science editor for BuzzFeed UK

"Ever since my first encounter with Professor Philip Moriarty (via a Sixty Symbols video), I have found his ability to explain and simplify complicated subjects rather beguiling. Surely someone as under-qualified as I could never grasp even a small percentage of the concepts that explain and predict the nature of the universe? But it appears he's made that possible. Once I'd met him, it became clear how he does it. His boundless enthusiasm makes you feel like a passenger on a luxury cruise, yet by the time he's finished his explanation you realize you've run a marathon alongside him. Witnessing his own excitement at the propagation of knowledge is just as rewarding as being the recipient of the knowledge. The author's energy, unique style, and ability to educate are not lost in print. From the first chapter, this book captures the imagination, effortlessly intertwining the fundamentals of rock music with the principles of quantum physics. His unapologetic inclusion of some (but not too much) math adds clarity without hindering the narrative. Whether you're a physicist, science enthusiast, musician, or music fan this book will entertain and enlighten in equal amounts. It will bring a new beauty to your favorite songs, and arm you with fresh concepts to explain some of the most counter-intuitive of scientific ideas. At the very least, you'll have an interesting conversational tangent to adopt next time someone wants to force their amateur rendition of 'Smoke on the Water' upon you. It's ironic that this book is about uncertainty, considering very soon I expect to be able to measure both its position in the book charts and the speed at which it arrived simultaneously. By rights it should be very high and very fast—deservedly so."

—DAVID DOMMINNEY FOWLER, guitarist with the
Australian Pink Floyd Show

WHEN THE
UNCERTAINTY
PRINCIPLE
GOES TO 11

WHEN THE UNCERTAINTY PRINCIPLE GOES TO 11

OR HOW TO EXPLAIN QUANTUM PHYSICS WITH HEAVY METAL

PHILIP MORIARTY
ILLUSTRATIONS BY PETE MCPARTLAN

BenBella

BenBella Books, Inc.
Dallas, TX

BENBELLA

BenBella Books, Inc.
10440 N. Central Expressway, Suite 800
Dallas, TX 75231
www.benbellabooks.com
Send feedback to feedback@benbellabooks.com

Printed in the United States of America
10 9 8 7 6 5 4 3 2

Library of Congress Cataloging-in-Publication Data is available upon request.
ISBN 9781944648527 (trade paper)
ISBN 9781944648534 (electronic)

Editing by Alexa Stevenson and Laurel Leigh
Copyediting by Scott Calamar
Proofreading by Michael Fedison and Sarah Vostok
Indexing by Clive Pyne
Text design and composition by Silver Feather Design
Cover illustration by Pete McPartlan
Cover design by Sarah Avinger
Printed by Versa Press

Distributed to the trade by Two Rivers Distribution, an Ingram brand
www.tworiversdistribution.com

Special discounts for bulk sales (minimum of 25 copies) are available.
Please contact Aida Herrera at aida@benbellabooks.com.

To Niamh, Saoirse, and Fiachra, who have patiently endured their father's appalling attempts to sing along with metal classics over the years . . .

CONTENTS

Chapter 1

PERMANENT WAVES

Wheels within wheels, a spiral array
A pattern so grand and complex

—from Rush's "Natural Science"[1]

It starts with a primal thud. A universal heartbeat.

Seconds later, a wall of sound explodes.

Through the mist, arms outstretched, a figure emerges. Guttural grunts and growls give way to the most haunting of melodies. The music builds majestically, spellbinding in its intensity, a complex soundscape underpinned by a deep emotional charge. The crowd becomes a choir, voices resonant with those onstage . . .

There's nothing quite like the sense-bludgeoning experience of a heavy metal gig. The all-enveloping power of the music, the theatrics, the histrionics . . . and the physics. Yes, the *physics*. Believe it or not, the links between heavy metal and quantum physics are especially deep and simply have not received anything like the attention they deserve. Quantum physics—also known as "quantum mechanics" or simply "quantum theory," because why have one name for something when you could have three, right?—is the physics of the invisible, the science of particles that are smaller than small. It's also in essence a theory of waves, and therefore the connections with the physics of music are already strong. But the stylings of heavy metal take these connections to another level entirely: chugging guitars, choked cymbals, artificial harmonics, and mosh pits each have their own parallels within the physics of the ultrasmall.[2]

---

[2] Just like physics, metal has a language and lexicon all its own. I've already slipped into the jargon, so let me briefly explain some of these terms for the uninitiated. A guitar "chugs" when the strings are muted with the palm of the hand; a cymbal is "choked" when it's grabbed to mute the sound; "artificial harmonics" are a way of making a guitar squeal by effectively pinching a string (hence the alternative term "pinch harmonics"); "mosh pits" are . . . areas near the stage that intrepid fans enter at their own risk. We'll come back to mosh pits.

I think it's safe to say that quantum physics has a reputation for being conceptually challenging. On the other hand, heavy metal—and its myriad thrash-power-sludge-stoner-hair-glam-death-progressive-djent-industrial-[*complete according to taste*] subgenres and subcultures—is not, it has to be said, generally considered to be the most cerebral of musical forms. Unfairly stereotyped as music for Neanderthals, frequently seen as the root of all evil (and, as such, a convenient scapegoat for societal problems whose origins are a great deal more complex than the lyrics of the latest Judas Priest album), metal is nonetheless often harmonically rich, lyrically challenging, and rhythmically complex.

. . . and, yes, I have to grudgingly admit, just as often it's not. But for every KISS, Mötley Crüe, or Whitesnake the metal critic will cite to make their case,[3] I'll counter with Opeth, Meshuggah, and Dream Theater. Then I'll raise the stakes with Mastodon, TesseracT, Queensrÿche, and Tool. And to clinch the deal, I'll close with Rush.[4] Each of these bands composes intelligent, intricate, and thoughtful music, their orchestrations frequently designed with what can only be called mathematical precision. (Indeed, there's an entire genre of metal known as math metal that prides itself on complex time signatures and "out there" arrangements.)

Although quite a number of bands have used scientific and/or mathematical themes as inspiration for their music—the super-talented and innovative Devin Townsend even titled one of his albums *Physicist*—that's not what this book is about. (I'll certainly be making more than passing reference to tracks and albums with strong lyrical/narrative links to science, however.) Nor am I preaching only to the converted. I love

---

[3] Astoundingly, with their 1984 album *Slide It In*, the 'Snake managed to out-Tap Spinal Tap—no mean feat.

[4] Opinion is divided as to whether Rush has ever been a heavy metal band. I'm a major fan of the Toronto trio and it will take a lot to convince me that Rush wasn't metal up until at least their sixth album. Let's consider the evidence: (1) Blues-based heavy riffs? Tick. (2) Virtuoso guitar solos? Tick. (3) Vocals in the stratosphere? Tick. (4) Long, flowing kimonos . . . Errmm. Okay, maybe not the kimonos. But otherwise, metal. Very metal.

metal. And I love physics. And I know from the logos that adorn the T-shirts of many physics students and researchers that I'm certainly not alone in this. But while this book has, of course, been written with those metal-loving physicists (or physics-loving metalheads, if you prefer) very much in mind, it's not just my "tribe" I'm hoping to connect with. My key motivation in writing this is to bring the beauty of quantum physics to a wider audience via the medium of metal. As we'll see, metal music is perfectly placed when it comes to crossing that age-old (and very silly) divide between the arts/humanities and the sciences. Each time your favorite band launches into *that* riff or *that* rhythm or *that* drum pattern, they're exploiting the very same principles of physics and mathematics that underpin how atoms, molecules, and quantum particles behave.

You'll notice that I slipped "mathematics" into the preceding sentence. I make absolutely no apologies in telling you that we're not going to go out of our way to avoid maths as we explore all of those fascinating quantum-metal parallels.[5] Some editors might claim that this statement alone would be enough to reduce a book's readership by 50 percent (or some similarly alarming figure no doubt plucked from thin air).[6] However, I have a great deal of confidence in the intellect and tenacity of the average metal fan (and the average reader in general).[7] Anyway, in all conscience, I can't drop the maths—it's the language of physics.

So: *there will be maths.* But contrary to popular belief in some quarters—I'm looking at some of you unreconstructed theorists here—physics is not *just* mathematics. And while to a mathematician, equations and functions have an elegance and a beauty all their own, for physicists it's the "unreasonable effectiveness" of mathematics in

---

[5] In the United Kingdom and Ireland, we say "maths." In the United States, it's "math." Potato, po-tah-toe. For authenticity's sake, I will stick to my native dialect.

[6] But most definitely not those I worked with on this book.

[7] One *Guardian* author quite agrees. See Ian Winwood's March 2007 post, "Why metal fans are brainier": www.theguardian.com/music/musicblog/2007/mar/21/whymetalfansarebrainier.

describing the world around us that never fails to impress.[8] The fact that so much of the behavior and structure of our universe can be captured by maths is truly remarkable. And that's exactly what we're going to see time and again in this book: the uncanny, staggeringly "unreasonable effectiveness" with which mathematics explains everything from the crunchiest of riffs and heaviest of rhythms to the far-beyond-driven vibrations of atoms and molecules.

A number of years back, I worked with a very talented musician called Dave Brown (aka the YouTuber Boyinaband) on a metal song whose riffs, rhythms, and, um, rlyrics (© D. Brown) were derived from the fundamental constant known as the golden ratio.[9] We uploaded a YouTube video for this math-metal mash-up and,[10] foolishly ignoring the wise counsel of friends and colleagues regarding the quality of online critique,[11] took a look at the comments. I was absolutely delighted to read the following:

Christina ▓▓▓▓▓ 4 years ago
I think you just tricked me into liking math.. Clever bastards.

REPLY 👍 👎

---

[8] Eugene Wigner, a Hungarian-American theoretical physicist who, among many other achievements, made far-reaching contributions to the development of quantum mechanics and played a key role in the Manhattan Project, wrote an influential (and, in some circles, somewhat contentious) paper entitled "The Unreasonable Effectiveness of Mathematics in the Physical Sciences," published in *Communications on Pure and Applied Mathematics* back in 1960. (At the time of this writing, a pdf of the paper is available online and is easily locatable via a search for the title, or see Wigner, Eugene. "The Unreasonable Effectiveness of Mathematics in the Physical Sciences." *Communications on Pure and Applied Mathematics* 13 [1960]: 1–14).

[9] Yes, I know Tool did something similar with their 2001 album, *Lateralus*. See "Tau of Phi." YouTube video 3:13. Numberphile, November 19, 2012. www.youtube.com/watch?v=aiibxmqXV9M.

[10] It's here: "Golden Ratio Song." YouTube video, 3:07. Numberphile, July 13, 2012. www.youtube.com/watch?v=nBgQPSUTWVM.

[11] The vast majority of my colleagues, including, in particular, PhD and early career researchers, are firmly of the opinion that YouTube comments sections represent nothing more than the condensed collective stupidity of humanity. The wonderful xkcd wholeheartedly agrees: xkcd.com/202/.

Thank you, Christina! The ethos behind *When the Uncertainty Principle Goes to 11* is exactly this.

Now before we can start to really get our heads around the maths and physics underpinning metal, we need to address a deep and fundamental question. A question that cuts to the core of everything in this book: Just what is sound? Or, if you prefer: What is noise? And just how is it that we—to quote Anthrax quoting Public Enemy—bring the noise?

Surveying the Soundscape

What's happening at the most fundamental level when we hear something?

Short answer: a heck of a lot of physics, quite a bit of biochemistry, and some amazing psychology.[12]

The somewhat longer answer . . .

The sounds we hear every day are due, ultimately, to vibrations. And an excellent source of vibrations is an amplifier stack wound up to eleven. The cone of the loudspeaker vibrates, causing the surrounding air to be shaken up. More precisely, the air molecules are periodically pushed together and separated by the vibrations of the loudspeaker, forming a sound wave. This means that there are regular changes in the number of molecules packed into a given amount of space—in other words, the air *density* varies. If we had a powerful microscope that could image the individual molecules in the air—and microscopes now exist that can not only pick out individual molecules but can see inside them to discern their chemical architecture[13]—we'd in principle be able to see that density vary right down to the molecular level.

---

[12] The extent to which the brain processes the raw audio signals it receives via the ears is best demonstrated by this fascinating "audio illusion" from the Franklin Institute in Philadelphia: "An Audio Illusion—Your Brain." YouTube video, 0:52. Posted by DailyPunjabiMedia, July 4, 2014. www.youtube.com/watch?v=tG9HSvNPVKQ.

[13] These go under the name of scanning probe microscopes (SPMs) and form the bedrock of the type of research we do in our group at the University of

There's a problem with this strategy, however. There are millions upon billions upon trillions of molecules buzzing around down there, and if we could see every molecule, we'd be overwhelmed; it'd be a massive case of too much information. Luckily, there is a technique that allows us to visualize sound waves—to see sound—in a more instructive way. It's a process developed by physicist August Toepler in the 1860s that's known as Schlieren photography. It works by exploiting the refraction that occurs when light travels through regions of air that have different densities. That might sound somewhat esoteric, but you've seen this effect in action many times before: the shimmering heat haze that appears on roads on a hot day is due to light being bent by the difference in air density. In that case, it's not sound energy that's creating the density variation, but heat energy.

Here's what a Schlieren image of a sound wave would look like, drawn reasonably to scale:

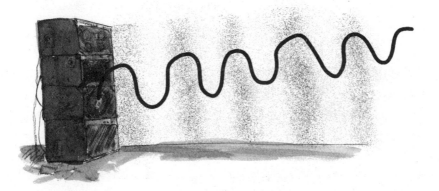

With Schlieren imaging, we can't see the individual molecules. But, nonetheless, we can directly observe the variation in the density of the air as the sound wave propagates. It's a remarkable technique.

Nottingham—we'll talk much more about these later. However, while imaging single molecules on a solid surface has been possible for decades, it's another matter entirely to attempt to see individual molecules in the air around us.

Sound is what's known as a longitudinal wave. All this means is that the movement of the medium—in this case, air—is *parallel* to the direction of motion of the wave. It's important to realize that the air itself does not travel with the wave—it's the disturbance that travels. In essence, the sound wave transmits energy, not matter. If you don't have an advanced Schlieren imaging system on hand (they don't come cheap), a Slinky can be used as a rather less costly—and, let's be honest, equally fun—tool to visualize a longitudinal wave. A Slinky in action mimics how a sound wave travels: the coils of the Slinky periodically move closer together (*compression*) and farther apart (*rarefaction*). The compression-rarefaction cycle of a Slinky's coils is analogous to the cyclic change in air density due to a sound wave:

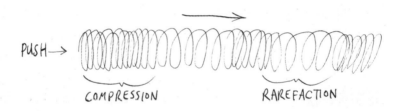

While we're playing with our Slinky, we can also generate the other type of wave motion that underpins so much of metal music: the transverse wave. In this case, the movement of the medium is *perpendicular* to the direction of the wave.

This type of wave is exceptionally important for our purposes because the ultimate origin of all those earth-shaking, teeth-rattling, and ear-piercing riffs and solos is the transverse wave that forms when a guitar string is hit with a pick. The transverse wave is in turn the origin of the longitudinal sound wave that reaches our ears; the motion of the string excites the air molecules,[14] and one type of wave motion is converted to the other.

We can use the wonderful audio application Audacity to capture the waveform produced by the string.[15] The illustration on the next page shows what Audacity displays for the opening note of Metallica's "Welcome Home (Sanitarium)." (Played on an acoustic guitar. This note, like most of the sound samples analyzed in this book, can be heard at the Uncertainty to 11 YouTube channel: https://www.you tube.com/channel/UCIg28nCrNa_gEHCEgPYfMqQ.) It shows just how the volume of the note—or, more correctly, its *amplitude*[16]—varies over time on two different timescales. The larger graph was made as the note rang out over about five and a half seconds. Although the gradual decay of the volume is clear—the note eventually dies away—it's difficult to discern any well-defined wave pattern on this longer timescale. If we zoom in on the time axis, however, and look

---

[14] . . . for an acoustic guitar. The conversion of "wave on a string" to "ear-piercing metal guitar solo" is more complicated for an electric guitar, involving conversion of a mechanical wave to an electrical wave and vice versa. Actually, it's complicated even for an acoustic guitar. The string vibration alone would not produce a very loud note. The body of the guitar is designed so that it resonates with the vibrating string. This leads to many more air molecules getting shunted back and forth, producing a much louder sound. Resonance will be a recurring theme throughout this book.

[15] I thoroughly recommend Audacity—it's a great piece of (free, at the time of this writing) software for recording, manipulating, and mixing sound samples, and I've used it a great deal not only for this book but throughout my undergraduate lectures and videos.

[16] Don't worry about the distinction between volume and amplitude—it's a subtlety we can ignore for the purposes of this book.

at what's happening on the scale of 100-milliseconds,[17] it's a different story indeed. The inset shows a very regular, periodic pattern as the guitar string cycles back and forth, driving the air molecules (which in turn drive the microphone and enable the signal to be captured by my laptop).

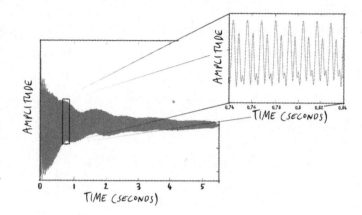

That sample makes it very clear that the waveform is cyclical—it repeats with a well-defined *period*. It's traditional in physics to use an uppercase T to represent this period of oscillation. And there's an exceptionally simple relationship between the period, T, and frequency, f, of the wave: frequency is the number of cycles the wave completes every second, so $f = 1/T$. We use the unit Hz, short for hertz, for frequency. The higher the frequency, the higher the pitch we hear. This simple relationship between pitch and frequency is often illustrated using a diagram something like this:

---

[17] A brief note of explanation about milliseconds and decimal places: 0.001 seconds is 1 millisecond (i.e., a thousandth of a second), 0.01 seconds is 10 milliseconds, and 0.1 seconds is 100 milliseconds. The zeroes after the "1" are often left out but in professional science we tend to include them to demonstrate the precision of a measurement. So, 0.1 = 0.10 = 0.100 = 0.1000 = 100 milliseconds. Here, I'm not going to bother with trailing zeroes unless entirely necessary. It's a bit neater that way.

The notes toward the left-hand end of the piano keyboard have a longer period than those toward the right-hand end. This means that they have a lower frequency—in other words, fewer cycles are packed in per second.

I fully realize, however, that keyboards have long divided the metal community. In the early 1980s, and as a direct reaction against the synth-driven electropop that dominated the charts at the time, Iron Maiden proudly declared "No synthesizers or ulterior motives" on the cover of their *Piece of Mind* album. (Nonetheless, just a few years later Maiden made heavy use of guitar synths on their 1986 release, *Somewhere in Time*. And a decade or so down the line, Fear Factory was spearheading electropop-metal crossover with a crushing cover of Gary Numan's "Cars," featuring Numan himself.)

For those who prefer their metal old-school and untainted by new-fangled technology, let's convert that keyboard-centric diagram above to something a little more appropriate for the genre . . .

It's the same idea: down toward the left-hand end of the neck we have the low notes (from the perspective of the guitarist, that is, and if you can forgive the right-handed–centric view; apologies to those lefties who are reading). At the other end of the neck we find the face-melting, eardrum-damaging, gurn-generating high-pitched notes that are the bedrock of the metal guitar solo. It's again just a simple matter of low-frequency vs high-frequency notes.

Or is it?

Here's what the opening note of "Sanitarium" looks like in standard musical notation for a guitarist:[18]

. . . And here's what it looks like for a pianist:

Spot the difference?

No?

---

[18] Don't worry if you have no experience reading standard musical notation. You're in good company—the vast majority of rock musicians don't read music either. And you're not going to need to be able to read music to understand any of this book.

That's because there isn't one.

Yet if I were to play that opening note on a guitar and then on a piano, you'd readily discern a difference—it would be clear that a different instrument had been used in each case. Why is this? After all, it's an E note regardless of whether it's played on a guitar, piano, flute, or kazoo. The frequency of that E note is 84 Hz in each case. So how do we instinctively know that the note has been played on different instruments?

Enter Fourier.

Pitches and Patterns

Let's travel back to a time long Before Sabbath (BS)—before the first distorted notes were wrung from an electric guitar, before rock and roll emerged from the blues, long before the blues itself arose in the Deep South. We're going back to the eighteenth century, to consider the remarkable insights and true genius of Jean-Baptiste Joseph Fourier. It's no exaggeration to say that Fourier radically changed the way we understand the world around us, on scales ranging from the subatomic to the ninety-three-billion-light-year diameter of the observable universe. And before we can understand the relationships between metal and quantum physics, we're going to need to take a look at Fourier's elegant approach to the analysis of waves and patterns.

JEAN-BAPTISTE JOSEPH FOURIER

Yesterday was my twenty-first birthday. At that age Newton and Pascal had already acquired many claims to immortality.

That's how the young Joseph Fourier voiced his concerns about his potential impact on physics and maths in a letter sent in March 1789 to C. L. Bonard, professor of mathematics at the university at Auxerre, near Paris. Fourier was writing from the Benedictine abbey at Saint-Benoît-sur-Loire, where he was training for the priesthood (with, it must be said, hardly unalloyed enthusiasm).

Fourier needn't have worried. The methods of mathematical analysis he went on to develop are now so thoroughly embedded in physicists' thinking that every physics undergraduate—along with students in a significant number of other disciplines, including engineering—encounters Fourier techniques many times over throughout their degree. Fourier's name echoes down through lecture theaters and conference halls, spanning generations. My erstwhile colleague at Nottingham, Peter Coles, an astronomer who at the time of this writing is head of Physics and Maths at the University of Sussex,[19] did his PhD thesis with the eminent cosmologist John Barrow and put it like this:

> This follows the best advice I ever got from my thesis advisor: "If you can't think of anything else to do, try Fourier-transforming everything."

It is certainly an approach *I've* found to be immensely helpful over the years.

Fourier's core idea is very simple to state: a pattern in space or time can be broken down into the waves that make it up. More specifically, Fourier analysis is a kind of translation that gives us a way of taking a complex mathematical function and breaking it into simpler functions. *That's it.*

Unfortunately, mention the words "Fourier analysis" to many who have completed a physics or engineering degree, and you'll provoke an involuntary shudder as long-suppressed memories of attempting to solve Fourier integrals come flooding back. It's a shame, but the revolutionary simplicity at the heart of Fourier analysis is often obscured by the (relatively) complicated mathematics involved in performing it. Fortunately, for our purpose, it's fairly easy to translate much of the

---

[19] I thoroughly recommend Peter's *In the Dark* blog (telescoper.wordpress.com) for entertaining and informative posts on topics spanning jazz to the large-scale structure of the universe.

mathematical notation into musical concepts.[20] (If you crave a more detailed explanation of the core mathematics, see the appendix on "The Maths of Metal" on page 311.)

For now, let's start with a demonstration of Fourier's methods in action. We're going to whistle a note. Yes, I know that whistling isn't very metal, but the great thing about a whistled note is that it's a simple, pure tone and, as is so often the case in physics (and is *especially* the case when it comes to Fourier analysis), we gain a lot of understanding by breaking a system or phenomenon down to its most basic elements. Perhaps the best example of whistling in hard rock, if not metal, is the start of Guns N' Roses' "Patience."[21] Axl Rose opens with a whistled A♯ (A-sharp) note.[22] Here's what my version of that whistled note looks like, recorded (as ever) with Audacity and observed over the course of 10 milliseconds:

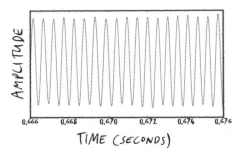

TIME (SECONDS)

---

[20] Mathematics is a language, and, as with any language, it can be translated into other forms. We might lose something in translation, but we can get the core ideas across.

[21] Another example from the hard rock canon, albeit in full-on-syrupy-power-ballad mode, is Scorpions' "Wind of Change." That particular track is outside the remit of this book, however, purely on grounds of taste and decency. (I have *some* standards.) Queensrÿche, who, in their prime, many would class as key exponents of prog metal, have also included whistling in both "The Lady Wore Black" and "I Will Remember."

[22] This, too, is somewhat of a "Potato, po-tah-toe" thing. If you want to plumb the depths of their nerdery, gather your musician friends sometime and ask whether a certain note is a B♭ or an A♯. (They are the same, in frequency; the definition generally depends on the key the song is in.) As the guitars are likely tuned down half a step on that recording, one friend has ventured that Axl is whistling over a Gb major pentatonic scale—an appropriately metal name for a scale if ever there was one.

It's worth comparing this waveform with the graph a few pages back of that E note at the start of "Sanitarium" as played on guitar. Go ahead, do it. The frequency is different—the E at the start of "Sanitarium" is a much lower pitch than the A♯ note whistled by Axl Rose (and that's why the timescales on the corresponding graphs are different)—but a significantly more important difference is that the *shape* of the wave-form is much simpler. The whistled note depicted above looks almost exactly like a "textbook" wave; it varies smoothly in a way that's straight out of Wave Physics 101. A physicist looking at the sample of that whis-tled note would say that it's very close in form to a pure sine wave. On the other hand, the opening note of "Sanitarium," whether played on guitar or piano, looks far more complicated.

I casually threw in the term "sine wave" there, but what in the name of Dio is a sine wave? Sine waves pop up everywhere in nature/science/ the universe because they're the natural solution to the mathematical equations that describe the motion of so many objects. I know that this may bring back long-repressed memories of trig class for many of you, for which I apologize, but we're going to have to take a look at the mathematical machinery underpinning a sine wave. This is of key importance because, among its many other fascinating properties, the humble sine wave is the cornerstone of Fourier's approach to analyzing patterns (of any form). To continue our quest to connect the likes of Quiet Riot and Queensrÿche with the quantum, we'll need to spend some time looking at the maths and the science of sines.

Chapter 2

BANGING A DIFFERENT DRUM

*To me, drum soloing is like doing a marathon
and solving equations at the same time.*

—Neil Peart[1]

---

[1] Interview with *MusicRadar*, 2011.

In order to explain a sine wave, I'm going to switch to an instrument we've not yet considered—the drums—and to a component of that instrument especially important in metal music, namely the bass drum beater. Double bass (DB) drumming is an integral component of so much of metal, ranging from the superhuman DB pummeling that powers death/black metal to the slightly less frantic, though nonetheless extremely effective, syncopated patterns driving the likes of Megadeth, Judas Priest, and Machine Head. And the core of DB dynamics is the spring-loaded motion of the bass drum pedal.

If we grab the drum beater and release it, it'll bob to and fro due to the action of the spring—as can be seen in the following superimposed frame(s), taken from a video of the motion of the beater. The beater is effectively a pendulum, rocking back and forth at its *natural* or *resonant* frequency. Make the spring stiffer, or use a heavier (or lighter) beater, and this frequency will change accordingly.

The energy we stored in the spring when we pulled back the drum beater (*potential* energy) is converted to energy of motion (*kinetic* energy). If we lived in the idealized conceptual world of spherical cows and frictionless surfaces that physicists so like to inhabit, this motion would continue indefinitely. With no damping—that is, no dissipation of energy (and we'll have lots to say about energy throughout this book[2])—the beater would swing forward and backward, always returning to the same position at which we kicked off the oscillation, until the end of time. In the real world, however, the beater gradually comes to a stop—due to friction in the pedal, the (small amount of) heat energy generated in the spring as it is alternately compressed and stretched, and even, to a certain extent, resistance from the surrounding air.

This conversion of "stored" potential energy to the kinetic energy that drives motion happens all the time in the world around us. Here's a very simple experiment to demonstrate potential-to-kinetic energy in your home. Walk over to your music collection, select a CD or DVD (or, if you're lucky enough to own vinyl, a record) that you're not especially proud to own—we all have *those* albums—and hold it at arm's length. Now, drop it.[3]

Before you dropped the album, it had what we call gravitational potential energy, simply by virtue of your holding it at a (small) height above the Earth's surface. We can even work out how much potential energy it had through a very simple formula (which you might recognize from secondary/high school science classes):

$$E_{pot} = mgh$$

In this formula m represents the mass of the CD and its case (about 85 g in total), g is the value of the acceleration due to the gravitational

---

[2] There are very few questions in physics (or, indeed, in all of science) that can't be glibly answered with, "Well, it's because the system wants to reach its lowest energy." The devil, of course, is in the details . . .

[3] If you don't own a physical CD/DVD/vinyl collection and all of your music is instead in cyberspace and accessed via, for example, an iPod or iPhone, don't worry, you can still do this experiment. Simply hold your iPhone at a height of 1.5 meters above a suitably hard surface . . . and let go.

force (which is [very] roughly 10 meters per second per second[4]), and *h* is the height of the CD above the floor (let's say 1.5 meters).[5] Plug in the numbers and we find that the value of E_{pot} is about 1.3 joules. (A joule, abbreviated J, is a measure of energy in physics.) Okay, fine, you might say, but how much energy does that actually represent in terms of anything I encounter on a day-to-day basis?

When attempting to place an amount of energy in an everyday context, scientists will often turn to the iconic Mars bar for their base unit, using the value of the nutritional energy available in a Mars bar to make a comparison with other less "tangible" forms of energy. Despite the multiplicity of forms of energy out there, all can be represented in terms of joules. But your CD's 1.3 J of gravitational potential is an *exceptionally* tiny amount compared to the 964,000 joules (or 230 kilocalories;[6] one calorie is the same as 4.2 joules) provided by a Mars bar.[7] Moreover, 964 kilojoules is only the "nutritional" energy value. If we could somehow directly capture the mass of the Mars bar as its associated energy, according to Einstein's famous $E = mc^2$, then we'd have a whopping 4,600,000,000,000,000 joules, or in scientific notation, 4.6×10^{15}. (In Einstein's formula, *m* represents the mass of the object, and *c* is the speed

---

[4] At first glance, "per second per second" seems like a rather strange expression. Here we're talking about an acceleration rather than a speed. Speed is the distance traveled in a certain amount of time and so is measured in meters per second. But acceleration is the rate of *change* of speed, namely the amount speed changes per second: (meters per second) per second.

[5] Strictly, the equation tells us the *change* in potential energy for the album at a height of 1.5 m compared to when it's on the floor—that is, the difference in potential energy due to gravity at the Earth's surface and 1.5 m above the Earth's surface (Only changes in potential energy matter).

[6] When we're told that we should typically consume between 1,500 and 2,500 calories a day, what's actually meant is kilocalories. So, um, 1 "calorie" = 1 kilocalorie. This, I think we can all agree, is unnecessarily confusing. And while we're on the subject of calorie intake, it would be entirely remiss of me not to mention the Vegan Black Metal Chef, who grunts and growls his recipes over a black metal backing track and whose YouTube videos have been exceptionally popular. See https://www.youtube.com/watch?v=CeZlih4DDNg.

[7] Source: the back of a Mars bar wrapper. Let no one say that I didn't do my research for this book.

of light. The speed of light is very fast indeed, about a billion kilometers an hour—or about 670 million miles per hour in old money—so even a small mass is associated with a very large amount of energy.[8])

Now, 1.3 J of energy is *peanuts* compared to this. Heck, even calling it peanuts overstates the case; the nutritional energy content of a single peanut is about 3 "calories"—3 kilocalories in proper units, which is about 12,500 joules. (The peanuts comparison does, however, allow me to crowbar in a reference to the wonderful heavy metal *Peanuts* video available on YouTube—at least at the time of writing.[9] If you haven't seen it, I strongly suggest that you put this book down right now and go take a look. Charlie Brown, Snoopy, and the rest of the *Peanuts* gang enthusiastically rocking out to some metal madness will definitely brighten up your day.) Before I wander too far off topic (into the wildly contentious area of nut-influenced metal; trust me, we really don't want to go there), the point of the Mars-and-peanuts diversion was simply to flag that there are many forms of energy and that we can readily convert between them. Now, let's get back to that beater.

The motion of the drum beater is a great example of the cyclic conversion of potential energy (stored by the spring) to kinetic energy (the beater in motion), and back again. At the extremes of the swing of the beater it's all potential energy and no kinetic energy, while at the center point, when the beater is vertical, the situation is reversed and it's all

---

[8] There is a key distinction between matter and mass. There are also very subtle and important differences between what's known as rest mass and relativistic mass (with usage of the latter term having fallen out of favor with some groups of physicists). $E = mc^2$ is strictly only true for a stationary object; for moving objects (and objects moving relative to each other), it gets a little—no, make that *a lot*—more complicated. I'm not going to spend any more time, however, on the distinctions because, as we scientists are fond of saying in our research papers, it's outside the scope of this work. I'd prefer not to go down the deep, and topologically complex, rabbit hole of relativistic physics here. I instead enthusiastically recommend Matt Strassler's *Of Particular Significance* blog (profmattstrassler.com) for much, much more on these important conceptual distinctions between matter, mass, and energy.

[9] "A Charlie Brown Heavy Metal Christmas." YouTube video, 2:24. Posted by Michael Massey, March 19, 2006. www.youtube.com/watch?v=4AC3sZB-v7Q.

kinetic energy. The word "kinetic" has its origins in the Greek *kinētikos*, which means "of motion." You therefore might not be surprised to find that kinetic energy and speed are closely related. In fact, the relationship between the two is very simple:

$$E_{kin} = \tfrac{1}{2}\, mv^2$$

E_{kin} represents the amount of kinetic energy, m is the mass of the object (in this case, the drum beater), and v is the speed at which it is moving.[10]

Although the equation is indeed simple and straightforward to use—multiply the mass by the velocity squared and divide by two—there's something there that you might find puzzling. Namely, why do we use the square of the velocity, v^2, and not just the velocity, v, to work out how much kinetic energy we have? If you're bemused by this, I've got two things to say. First, it's good that you are! No, really, it is. You're thinking like a scientist: skeptical, critical, not wanting to take statements at face value. And secondly, you're not at all alone in being perplexed by this—it was a source of great confusion and debate among scientists for many years both in the early days of what we now call physics and during the development of calculus. (Look up the Wikipedia page on "vis viva," for example.) In the appendix, I explain just why it's v^2, rather than v, but for now, all you need to understand is that although the kinetic energy of an object depends on its speed, the relationship isn't linear. Double the speed and the kinetic energy doesn't just increase by a factor of two, it goes up fourfold.

---

[10] There's an important difference between *velocity* and *speed*. Velocity is what's known as a vector quantity in physics. Vectors have both a size and a direction. So, "We were moving due north at about 60 km/hr, Officer, when Vic Skullcrusher, our guitarist, stepped out of the door of the tour bus, having mistaken it for the toilet" is an example of velocity, i.e., speed in a particular direction. (And, yes, that particular Spinal Tap–esque incident really happened to a UK band. Names have been changed to protect the guilty.) When determining kinetic energy, however, we only care about the magnitude of the velocity: the speed, and not the direction.

The equation for the potential energy associated with the spring of the drum pedal, which we'll call E_{pot}, is also rather straightforward:

$$E_{pot} = ½\, kx^2$$

Here, k represents the stiffness of the spring and x is how much we've compressed or stretched the spring. As you might expect, the stiffer the spring, the more energy that can be stored. That's just because we have to put in more work to compress (or stretch) that spring in the first place. (We can't beat the first law of thermodynamics: energy is conserved.) But note that, just like the dependence of kinetic energy on velocity, the potential energy scales with the *square* of the distance by which we've compressed or stretched the spring.

So, at this point we've established that our beater has potential energy and kinetic energy, and that one form of energy will be converted to the other. Now let's ask what type of motion this interconversion of energy will produce. You might correctly guess that it'll be cyclic, but just what trajectory will the beater follow? The answer is at the very core of everything we'll do in this book, and it's the basic ingredient of Fourier's approach to analyzing patterns.

We could continue to explore this from a purely theoretical perspective (and, indeed, in the appendix on "The Maths of Metal," I discuss how we can derive a formula for the motion of the beater from a consideration of the energy flow). But I'm a dyed-in-the-wool experimentalist; data is everything. So let's do the experiment. It's more fun that way in any case.

Let There Be Drums. And There Was Drums . . .[11]

In order to track the motion of the drum beater with some degree of accuracy, I set up a simple tabletop experiment in my office. Although

---

[11] AC/DC, "Let There Be Rock," 1977.

far from the state of the art in motion-detection technology, this Heath Robinson approach proved more than good enough to measure the trajectory of our drum beater.[12] I first stuck a small mirror onto the end of the beater. Then I bounced the beam from a laser pointer off that mirror onto the wall, set the beater in motion, and filmed the resulting movement of the laser spot. The setup looked like this:

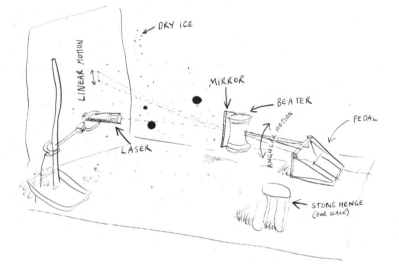

📹 You can see the experiment in action on the Uncertainty to 11 YouTube channel: https://www.youtube.com/watch?v=1mqITu-uU90.

We'll identify the drum beater's position as its *angular* position—in this case, the angle between it and the vertical axis. Here's what we get when we plot the position of the beater at different instants of time . . .

---

[12] William Heath Robinson was a British cartoonist and illustrator known for his fascinating sketches of over-the-top machines designed to perform very simple tasks. I first encountered the phrase "It's a bit Heath Robinson" when I started my postdoctoral career in Nottingham, was puzzled by it, and did some research into the man. Intriguingly, Heath Robinson is now generally associated with temporary solutions to engineering problems, or prototypes, that are far from sophisticated but get the job done with whatever bits and bobs might be on hand, including tape, rubber bands, string, etc.

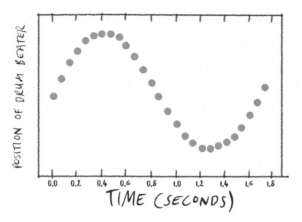

You may well recognize the shape formed by the data points. But what we really want to know is how we can *quantify* that shape—how can we write a formula that describes that shape? Or, in more mathematical language, what *function* does the pattern of dots represent? Can we find a mathematical way of relating the position of the beater over time that accurately represents the data we have?

This is physics in action—this is how we physicists analyze data. I should stress that, as an experimental physicist, empirical data are, and always will be, sacrosanct to me. As Richard Feynman put it so well:[13]

> *If it disagrees with experiment it is wrong. In that simple statement is the key to science. It does not make any difference how beautiful your guess is. It does not make any difference how smart you are, who made the guess, or what his name is*[14]*—if it disagrees with experiment it is wrong. That is all there is to it.*

---

[13] Richard Feynman (1918–1988) is not only the archetypal physicist's physicist but was remarkably accomplished at communicating science to laypeople. Long before public engagement was seen as the "done thing" in academic circles— thankfully, we live in more enlightened times—Feynman communicated his love of physics to a very broad audience. As such, it is now verboten not to mention Feynman in pop physics books like this.

[14] Female physicists of note include Kathleen Lonsdale (1903–1971), who was the first to apply Fourier's methods to the analysis of the structure of molecules.

So let's do this analysis like good physicists. As Feynman suggests, we'll guess a mathematical function and see if that function does a good job of matching the pattern we see in our observation of the drum beater. We're looking for a function that represents the value of the position of the drum beater, which we'll call θ, in line with physics tradition[15], that we should expect at a particular time, which we'll call *t*. We write this as θ (*t*) (in other words, "angular position as a function of time").

Here's an example of just about the simplest type of function we could try:

$$\theta\,(t) = 30°$$

This means that we're guessing that the angle is a *constant*; that at any moment of time, the beater will be found at an angle of 30 degrees. We could also write this as:

$$\theta\,(t) = \theta_c$$

Here we've replaced the 30 degrees with θ_c, representing the fact that the angle remains constant—that's what the little subscript *c* stands for. Of course, we could guess values other than 30 degrees for that constant angle, perhaps depending on our experimental setup. That's why the expression θ_c is useful—we can set the variable to whatever value is appropriate.

However, this function is clearly a very silly guess indeed because it flies in the face of our experimental observations. (Though physicists are not always immune from making silly guesses in their work.) The

---

[15] There are *lots* of confusing and frustratingly flexible traditions associated with the choice of symbols in physics. Theta, θ, is typically used for angle.* Or, errrm, temperature. Or a fundamental particle known as the pentaquark. And that's just physics. In soil science, θ represents water content! This Wikipedia article lays out the not-at-all confusing variety of things mathematicians and scientists use Greek letters to mean: https://en.wikipedia.org/wiki/Greek_letters_used_in_mathematics,_science,_and_engineering.

*Except when it's not, and we decide we'd like to use φ for angle instead.

beater is clearly moving, not remaining at a fixed angle! So that mathematical function is really a dreadful failure at capturing the beater's motion. What could we try next? Maybe we could suggest that the beater's position changes proportionally to the amount of time that has passed? We might try assuming further that it's a direct proportionality; that after, for example, twice the amount of time has passed, the angle has also increased by a factor of two. This would be what mathematicians, engineers, and scientists call a linear function, because if we were to make a graph of the angle vs. time, it would be a straight line.

Yet the graph of the beater's motion is clearly *not* a straight line, so that really doesn't get us much further, does it? We can see that the beater is swinging back and forth—it's regularly changing its direction. Therefore we can rule out a linear function as a good description of the dynamics of the drum beater; the angle of the beater to the vertical axis can't just increase continually—nor, for that matter, can it decrease continually. We need a function in which it will increase, then decrease, increase, then decrease, in a periodic manner. In other words, we can rule out all functions that don't *oscillate*.[16] We are seeking a mathematical description that takes into account that the beater moves regularly back and forth.

Now we're getting to the crux of the matter. There really aren't many mathematical functions that fit the bill when it comes to accurately capturing the motion of the beater, because there aren't many that oscillate in the way we need. We can speed things up a bit by asking a computer to find the best possible curve to represent our data. This process, called *fitting*, is used extensively in science to compare experimental results with

---

[16] This means that functions where the angle of the beater increases not just proportional to the time elapsed, but to the square of the time elapsed (t^2) or the cube (t^3), or any power, don't work. Nor does a function in which the angle changes exponentially, or inversely (i.e., $1/t$) with time. And that's because none of those functions oscillate. (Although if you check out the appendix, you'll find that there's a very close relationship between an exponential function and a sine function under the appropriate, complex conditions.)

predictions from theory, checking data generated by a mathematical/computer model against real-world measurements.

Here are the measurements of the position of the drum beater again, this time with what's called the "best fit" curve superimposed as a solid line . . .

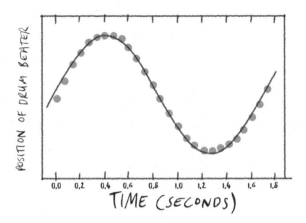

The curve that the computer has fitted corresponds to a mathematical function. And, as you may have guessed, in this case that function is a sine wave. As the beater moves back and forth, it traces out a sine curve.

Mathematically, we can describe the motion of the beater as follows:[17]

$$\theta(t) = A \sin(\omega t)$$

The function above is the core building block of Fourier's methods; it's at the very heart of the connections between metal and physics (and a wide variety of other phenomena as well). There are a couple of new terms (namely, A, ω, and sin) to explain here, but for those of you worrying that this book is going to turn into an undergraduate physics textbook, rest easy—it's really not going to get much more mathematical than this. (And for those of you who prefer the maths in all its gory detail, there's always the appendix.)

---

[17] This is just one way of writing the function. We'll see soon that there are other ways of doing this.

Let's break down the mathematical language of that function and explain it in English. First of all, on the left-hand side of the equal sign we have $\theta(t)$. You already know what that means: the angle the beater makes with the vertical (its angular position) depends on (is a function of) time. But it's the right-hand side of the equation that tells us just *how* the beater's position depends on time. The "sin" is shorthand for "sine." (Yes, I know that just dropping one letter doesn't really shorten the word a great deal, but I didn't make the rules for mathematical nomenclature.) We can see that this sine function does a very good job of mathematically representing the periodic motion of the beater. And, indeed, sines pop up *all the time* when periodic motion is analyzed. But why?

First, let's explain what's meant by the sine of an angle with a suitably metalized example. Imagine the world's largest Marshall stack, as shown below. You're standing 30 meters back from the stage, and to see the top of the stack you need to strain your neck muscles (which have already taken a hammering from decades of head-banging) and look up at an angle of 60 degrees. The sine function is a ratio we can use to find distances when we have an angle, or an angle when we have distances. It is what allows you to calculate the height of the stack without having to ask a roadie for a tape measure (and a very tall ladder).

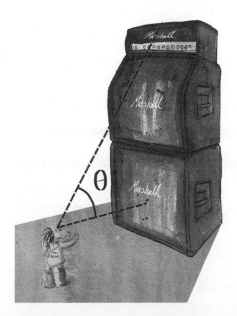

This type of basic trigonometry, for that's what it is, is reflected everywhere in the world around us—architects, engineers, scientists, and builders all exploit trigonometry, either consciously or unconsciously, in what they do. (It's worth noting that Pythagoras—whose [self-titled] theorem suggests he had more than a passing familiarity with triangles and angles himself—played a major role in establishing just where on a guitar the frets ought to be placed.)

The sine function, in essence, is a way of translating between distance, or position, and angle. To hammer this home, let's assume that the humongous Marshall stack we considered above is part of the backline for a suitably bombastic metal band, Metallizer,[18] playing their hometown arena. As in most arenas, the seats are laid out in blocks with rows designated by letters, and seats within those rows labeled with numbers. Let's say that you're seated in Row C, Seat 5, as shown below. We can represent your position in that block in two ways: with your x and y coordinates (where x is along the horizontal and y is along the vertical in the floor plan), or using the "as the crow flies" distance between your seat and the entrance to the block—which we call r in the sketch below[19]—along with your angular position.

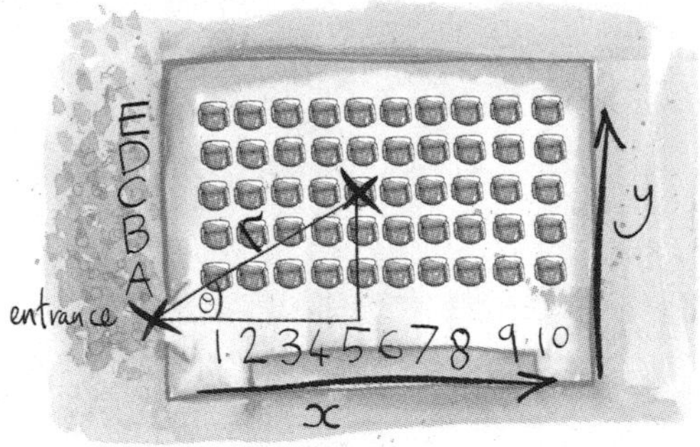

---

[18] Metallizer is, of course, not to be confused with the Brazilian band Metalizer, whose entry in *Encyclopaedia Metallum*, that repository of metal knowledge and wisdom, lists *The Thrashing Force* among their most recent albums.

[19] Note that "r" here is not a random choice of letter. It denotes that we're talking about a radial distance (the distance along the radius of a circle). We'll see the importance of circular motion very soon.

Translating between the two is what physicists call a switch of coordinate system. We've moved from a Cartesian (grid-like, or *x-y*) representation to what's known as a polar (*r*, θ) description of position.[20] And that type of switch is often essential—or, at the very least, extremely helpful—when tackling problems in physics, including that drum beater motion we're considering.

But back to the question at hand: we asked why the sine function pops up so often when analyzing periodic motion. To see the connection between the trigonometry of sines and periodicity, we're going to focus on the circle pit that breaks out when Metallizer are giving it their all midset.[21] (We're assuming that there's a standing-only area, as well as the seated blocks, otherwise it could get very messy indeed.) If we run in a circle, we eventually come back to where we started, so there's an inherent periodicity to our motion: round and round (as that '80s hair-metal classic goes). To keep things simple, let's consider the motion of just one of Metallizer's fans in the pit; let's also assume that she manages to make it around without bumping into other moshers—i.e., her path is unimpeded—and that her motion is a perfect circle.

The sketch on the next page shows how the trigonometry and periodicity of the sine function are entwined. The mosher—let's call her Marsha—starts at an angle of 0 degrees, and we've subsequently shown her at 45-degree intervals around the edge of the pit. We can use the sine function to work out her position along the *y* axis (just as in the seating plan example on the previous page). We set the center of the circle as (0,0); *y* positions north of this are positive, and *y* positions south of the center are negative. Similarly, the sine function switches from a positive to negative value when the angle increases above 180 degrees.

---

[20] Yes, we've used θ again here to represent angle, just as we did with the drum beater. In this case, θ is measured from the horizontal, whereas with the beater we measured from the vertical. Don't let this worry you. All that matters is the angular *displacement*—how much the angle changes. As long as we're consistent with our choice of the zero angle position in any given situation, we can move it around to best suit our needs.

[21] There's much more on the physics of metal crowd dynamics in Chapter 11.

Of course, just as Marsha the Mosher cycles through different positions, and different angles, to return back to where she started, so too must the sine function show this cyclic behavior. The sine of an angle has a value ranging from −1 to +1. This is clearly seen in the sketch, where the circle pit has been divided up into octants (and quadrants). If we look just at the quadrants, we see that at an angle of 0 degrees, the sine function has a value of 0; at 90 degrees, it's +1; at 180 degrees, we're back to 0 again; at 270 degrees the sine function is −1; and after one full revolution we're back to a sine of 0. Then the entire cycle repeats; round and round the mosher goes.

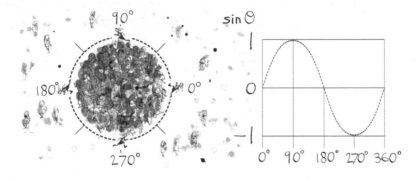

We're all very used to thinking of angles in terms of degrees; in the metal sphere (or should that be the metallosphere?), Def Leppard's "in the round" stage show for their *Hysteria* tour was touted as the first "full 360-degree experience," and many bands have been described as pulling off a "180-degree turn" when their music radically changes direction. But when you think about it, having 360 degrees comprise a full revolution is rather arbitrary. Why not 666 degrees, or 2,112 degrees, or any other randomly chosen number?

We can, in fact, trace the origin of the 360-degree figure back to the Babylonians—it's purely a historical artifact. But there's a much more mathematical (and, indeed, much more natural) way to define an angle that is used throughout science, maths, and engineering, and that doesn't rely on some arbitrary number of degrees in a circle. It's the *radian.* Our mosher friend traverses an angle of one radian when the distance she has run around the circle pit is the same as the radius of the circle (which we'll call *r*). It's an elegant concept and can be explained at a glance via the diagram below (where the circle has a radius of 1 meter). Move a distance *r* around the circumference of a circle and you'll have shifted your angular position by a radian. Simple as that.

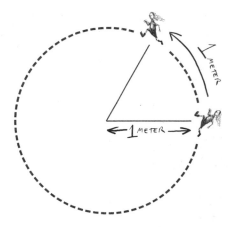

The circumference of a circle is given by the simple formula $2\pi r$ (where *r* again is the radius of the circle, and π is everyone's favorite ~~baked good~~ mathematical constant). So one revolution around the circle pit, or a 360-degree rotation, is an angle of 2π radians;[22] half a rotation is π radians; one-quarter of a rotation is $\pi/2$ radians, and so on. Not only is this more elegant mathematically (in that we're not arbitrarily slicing a circle into 360 parts), but when it comes to the calculus of trigonometric functions such as sine, radians make life a heck of a lot easier.

---

[22] There's been a lot of debate about whether π should be "superseded" by a fundamental constant known as τ, which equals 2π. You might think, like I did on first encountering the debate about tau vs pi, that this is rather silly mathematical pedantry but, from the perspective of maths teaching, there are certain advantages to thinking in terms of units of 2π rather than π. Brady Haran and I made a video about this for the Numberphile channel a few years ago: www.youtube.com/watch?v=83ofi_L6eAo.

The "unreasonable effectiveness" of mathematics in describing our universe again takes center stage when we compare the mosher's circle pit to the motion of the drum beater. What's remarkable is that the mosher's circular motion maps directly onto the movement of the beater. The same sine function describes both processes because, ultimately, *the motion of the drum beater is exactly the same as that of the mosher*, except that it's constrained to one axis. To see this, one need only take a look at the beater "face on," as in the sketch below. The position of the beater is basically a projection of the mosher's circular motion onto a single axis.

All that means is that if we plot the *y* position of the mosher vs. time, we get a graph that looks identical to the position of the beater vs. time: *because it's exactly the same mathematical function*.

Okay, so we've now got a working understanding of the sine function and how its trigonometry and periodicity are interrelated. Remember, however, that I was explaining the terms in this equation:

$$\theta(t) = A \sin(\omega t)$$

Note that, in this case, the "input" to the sine function—in other words, what determines the angle we're taking the sine *of*—is the product of the variables ω and t. You know that t represents time; but what is ω? This is simply the *natural* or *resonant* frequency of the beater I mentioned a few pages back: the beater moves to and fro at a particular rate, ω, which can be adjusted by varying either its mass or the stiffness of the spring (or both).

You might also quite reasonably ask why frequency is represented in this case by ω rather than f, as we had previously. That's a very good question. The difference is that ω represents an *angular* frequency, that is, the number of revolutions per second, measured in radians, whereas f is used to represent the number of cycles per second, measured in Hz. We're going to use both in this book, as is common throughout physics and engineering, because the distinction is very important. These two types of frequency are related as follows:

$$\omega = 2\pi f$$

There's that factor of 2π again, and it falls straight out of the discussion of radians above. A frequency of 10 Hz (10 cycles per second) means an angular frequency of 10 revolutions per second. We convert between the two by multiplying by 2π because it takes 2π radians to make up a single revolution. (Like a radian, a revolution is an angular measure—the angular counterpoint to the circumference of a circle, which, of course, is 2π multiplied by the *radius*.) For example, if our mosher races around the pit once in 2 seconds (a frequency of ½ of a cycle per second = 0.5 Hz), her angular frequency is 2π x 0.5 Hz = 1π radians per second. Or, alternatively, think of it like this: a single circuit of the pit takes 2 seconds. Therefore in 1 second the mosher slices an angle of $2\pi/2$ radians = π radians—half the circle.

We're now into the closing song of the set when it comes to deciphering that crucially important A sin (ωt) equation; all that's left to cover is what the A represents and we'll have finished dissecting this fundamental building block of Fourier's method of analysis (and thus have paved the way to an understanding of the metal-quantum nexus).

The A in the equation represents amplitude—the height or "strength" of the sine wave. The larger the value of A, the bigger the sine wave. As we've seen, the sine function only "outputs" values between −1 and +1, but our sine waves are not, of course, limited to these values. The A factor scales up the sine, accounting for a larger swing of the drum beater, or representing a wider circle pit for our—now totally exhausted—mosher, or signifying the increase in volume when the sound level for Metallizer's virtuoso guitarist is turned up in the mix.

Mention of sound levels brings us back at last to that whistled note we were considering before I embarked on this detailed, but mandatory, explanation of the sine function. (Thanks for bearing with me.) You'll remember that, unlike that of the ominous opening E of "Sanitarium," the waveform of the whistled note at the start of "Patience" was simple. It appeared to be a textbook sine wave, of just the sort we've seen describe the motion of both the drum beater and Marsha round the circle pit— there's that "unreasonable effectiveness" of maths again. But this time there's also a hugely important unifying *physical* principle accounting for the fact that both the drum beater's motion and the whistled note are so well described by a sine wave. And that principle is *resonance*.

Both the beater and the whistled note have a natural or resonant frequency: a frequency at which they naturally "want" to vibrate or oscillate. We're going to be seeing an awful lot of resonance in this book; it connects many different phenomena on wildly different length–scales in our universe, from the subnanoscopic world of quantum mechanics up to the patterns of orbits in planetary systems (with, of course, metal in the middle).

I included the scare quotes around "want" in the preceding paragraph because it can often be misleading when scientists anthropomorphize like that. There is, of course, nothing in the drum beater that "wants" to produce a particular frequency; it's obviously not sentient! Nonetheless, the particular physical characteristics of the system—in this case, the mass of the beater and the stiffness of the spring—establish the conditions for the system to have a maximum response at a particular frequency, or, more generally (and as we'll see in later chapters), to have a set of strong responses at a number of different, specific frequencies.

Resonance is core to physics, it's core to quantum mechanics, and it's core to metal. In the next chapter we'll see how Fourier's analysis not only provides deep insights into resonance, but dramatically simplifies how we think of waves in general—including that opening note of

"Sanitarium," which we'll finally get round to dissecting (literally). As a certain distinctively coiffed twentieth-century physicist allegedly, but most likely apocryphally, put it: "*Any intelligent fool* can make things bigger, more complex . . . It takes a touch of genius . . . to move in the opposite direction."

Let's move in the opposite direction.

Chapter 3

FROM FOURIER TO FEAR FACTORY

Mathematical analysis is as extensive as nature itself . . .
It brings together phenomena the most diverse, and
discovers the hidden analogies which unite them.

—Jean-Baptiste Joseph Fourier (1768–1830)

Algebra the way, never you betray
Life of math becoming clearer

—Internet parody of Metallica's *Master of Puppets*[1]

---

[1] As of this writing, you can find the jokester's lyrics at: www.amiright.com/
parody/80s/metallica64.shtml.

Master of Puppets is in my top ten albums of all time. Barely a week goes by when I don't listen to that metal masterpiece in its entirety. This is of course largely due to the arrangements, the production, the musicianship (yes, even Lars Ulrich's drumming[2]), and the vocals.

And, of course, the riffs.

Oh my, the riffs.

Cthulhu be praised, those *riffs*.

But I'd be somewhat disingenuous if I didn't admit that there's also a bit of nostalgia underpinning my love of *Puppets*. The first major metal concert I attended was the Dublin gig on Metallica's Damage Inc. tour for *Master of Puppets*.[3] James Hetfield had broken his wrist in a bizarre ~~gardening~~ skateboarding accident and so was solely on vocal duties, while his rhythm guitar parts were played with admirable precision by John Marshall, Hetfield's guitar tech at the time but also of Metal Church fame.[4] It was a hugely impressive gig: a band in their

---

[2] Ulrich gets an awful lot of stick for the quality of his stickmanship these days, and it indeed can be rather sloppy. But on *MOP* Ulrich did his fair share of inventive drumming that had a big impact at the time. It's important to consider the state of the art in the metal scene in 1986 to put Ulrich's drum tracks/arrangements in context. If we just take the "big four" of thrash metal—Metallica, Slayer, Anthrax, Megadeth—then Ulrich is indeed not the most technically accomplished of drummers in comparison. But advanced technique is not everything; inventiveness and ingenuity are equally, if not more, important. And, in any case, Ulrich did demonstrate technical prowess in spades on certain songs. It's just that he now seems unable or unwilling to re-create that in the live environment.

[3] September 14, 1986, at the St. Francis Xavier (SFX) Hall, Dublin. This was just twelve days before the tragic passing of Cliff Burton.

[4] Marshall also garners kudos from this particular Primus fan for being a one-time member of Blind Illusion, which featured the talents of Les Claypool and Larry LaLonde.

prime firing on all cylinders—Hetfield's wrist notwithstanding—and raised the bar well above that previously set by the new wave of British heavy metal (NWOBHM).

I hadn't, however, succumbed to Metallica's charms immediately. The first time I heard *Puppets,* a couple of months previously, I was underwhelmed, thinking it rather "samey." I was much more a fan of Anthrax—who opened for Metallica at that Dublin gig—due, in part, to the incredible vocal talents of their singer, Joey Belladonna. (Scott Ian, Anthrax's rhythm guitarist, also kept referencing the British comic *2000 AD* in interviews around that time. As I was a huge *2000 AD* fan, this couldn't help but endear Anthrax to me.)

That Dublin gig made me reconsider my opinion of Metallica, however. And something kept drawing me back to *Master of Puppets*—I found myself getting addicted to it. As a rookie guitarist playing in my first band,[5] one aspect of *Puppets* that made a decided impression on me was that massive guitar tone. It's still my favorite rhythm guitar sound on any metal album, including Metallica's later gazillion-selling *Black Album.*[6]

In Metallica's video diary of the making of the *Black Album*, there's a classic scene where Hetfield is laboring over his guitar tone by baffling and shrouding his amp in a "tent of doom" made up of, as he describes it, "foam and U-Haul blankets." When the sound damping is sorted to his satisfaction, Hetfield strikes an E chord. The resultant gut-thumping crunch from the amp causes him to break into a wide grin,

---

[5] We initially called ourselves Trauma, until someone pointed out that had also been the name of a band Cliff Burton had played in before he joined Metallica. Then we rather unimaginatively called ourselves Twist of Cain, because, um, we quite liked the Danzig song of the same name. We finally settled on Silent Witness as a moniker.

[6] Ty Tabor's sound on King's X's innovative, and shockingly overlooked, debut, *Out of the Silent Planet*, comes a very close second in my fave guitar tones. Opinion is divided as to how much influence King's X had on the burgeoning grunge movement at the time (the late[r] 1980s) but they were certainly among the front-runners in exploiting drop D tuning to get that heavy-as-lead guitar sound that later became synonymous with grunge.

with a guttural "ha ha." (It's only a shame that it wasn't his trademark "Yeah!") I am confident that this moment resonated deeply with every guitarist watching the video. Tone is everything.

From the signature Van Halen "warm brown" sound, through the scooped crunch of *Master of Puppets* that I love so much, to the set-the-gain-levels-to-kill assault of Dimebag Darrell and the phenomenal sustain of Nigel Tufnel, we all strive for that particular combination of amp and effects settings that will allow us to best re-create the sounds we hear in our heads.[7] Some guitarists even claim to be able to detect a difference in their sound depending on the type of battery in their effects pedals. I have my doubts about this—I'd like to see a reproducible result under well-controlled experimental conditions—but it's certainly the case that guitarists can be neurotic about their tone. Randy Rhoads's confidence both onstage and in the studio was directly influenced by the extent to which he could re-create his sound, and Rhoads was certainly no slouch on the fretboard.

We're going to see in this chapter and the next that control of guitar tone—including the tonal shaping provided by that staple of heavy metal soloing, the wah-wah pedal—is, at its core, an exercise in Fourier processing. (Indeed, any time you tweak the EQ on an iPod, a stereo, or an achingly state-of-the-art[8] 7.1 surround sound system, you're doing Fourier processing yourself.)

To preface our analysis of metalized and mangled guitar signals, let's revisit that whistling at the start of "Patience" once again. When we whistle, we shape the cavity in our mouth with our tongue to produce different notes; the air in our mouth resonates at a particular frequency set by that cavity shape. By blowing through our lips, we excite this resonance: we drive the air to resonate at a preferred frequency. It turns out that the physics of the humble whistle are quite complicated. Fortu-

---

[7] Many musicians now use purely digital processing via tools such as Guitar Rig and Axe-FX to shape their sound. Guitar Rig, in particular, has a large library of preset patches spanning a wide range of genres, and I've used it exclusively for all the guitar samples in this book.

[8] . . . at the time of this writing.

nately, all we need take note of is that when we whistle, we drive the air back and forth at a particular resonant frequency, which generates—as we saw—a pure sine wave, or close enough as makes no odds.

The graph inset in the figure below shows the whistled note just as we represented it in Chapter 1. We see that the sound of the whistle oscillates sinusoidally (er . . . as a sine wave) back and forth between a maximum and a minimum value. What's not clear from the graph, but can be easily measured, is that the peaks of the sine wave are separated by a little over half of a millisecond (0.534 ms, to be more precise). This is the *period* of the wave, which we represent mathematically with T. (We use a capital letter to distinguish the period, T, from bog-standard time, t. This distinction is illustrated in the graph below.) Period is generally expressed in terms of seconds, so here $T = .000534$ s.

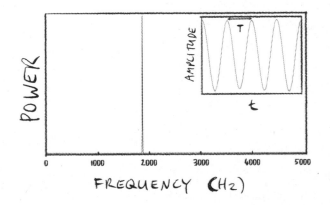

The frequency of the wave, as we discussed in Chapter 1, is simply $1/T$, i.e., the number of cycles per second. I will stress again that this reciprocal relationship between time and frequency is crucial. Fourier's method of analysis allows us to translate between the two: to move from the time domain to the frequency domain (and back again). That might sound like a rather esoteric mathematical statement, but it's a very simple principle: we can consider how a wave behaves as a function of time, as in the graph in the inset above, or we can instead choose to represent it in a much more compact way, as in the main graph. Each representation complements the other, and both are equally valid ways of viewing the wave: same wave, same data, different representation.

The main graph on the previous page shows the frequency *spectrum* of the wave: it's simply a way of representing the frequencies that are present in the wave. In this case it's a very simple spectrum: there is only one pure sine wave, so we see one spike in the spectrum at that frequency, 1,870 Hz.[9] The height of the spike tells us how much of that 1,870 Hz signal is present; in other words, it's an alternative way of representing the amplitude, or volume, of the whistled note. So, if we whistle more loudly, we'd expect the height of the spike to increase; whistle at a higher pitch and we'd suspect that the spike will shift along the x-axis of the graph to a higher frequency.

As ever, we should do the experiment to confirm our hypothesis. Here's how the frequency spectrum changed as I whistled the first three notes of "Patience" (I even put on a bandanna and leather shorts to channel Axl Rose when whistling[10]):

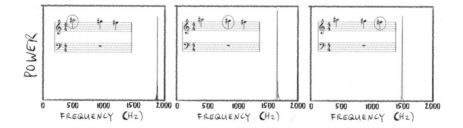

---

[9] The physicists and engineers among you may now be rushing to check the frequency of an A♯ note. In this case, it's an A♯6 note (where the 6 represents the octave for the note). The accepted reference frequency for that note is 1,865 Hz, so either Axl Rose or I was off by about 5 Hz. (We all know anything goes with Axl Rose, so I'm laying the blame squarely on his shoulders.)

[10] I didn't.

So there you have it. *That's Fourier analysis*. We've just done it—we translated a waveform or signal from a representation in one domain to a representation in a "reciprocal" domain, in this case from a representation in time to a representation in terms of frequency. If you've ever watched the bars of the display of a spectrum analyzer move up and down in time with the music, you were watching Fourier analysis in real time. Physicists even call the frequency spectrum the *Fourier spectrum*, and label the spikes Fourier components.

But if that's all there is to this Fourier lark, why, in the name of all that's south of heaven, does it terrify so many physics and engineering students, as was claimed back in Chapter 1? While it's a simple concept, the mathematical manipulations required to analyze all the various types of signals and waveforms that can crop up in physics (and engineering and chemistry and biology and medicine and so forth—the applications of Fourier analysis are virtually limitless) can get very hairy indeed. And complexity is introduced not only by more complex waves, but also by ostensibly simpler ones: our whistled note was *nearly* a perfect sine wave, but a *totally* pure sine wave—one of those conceptual idealizations physicists so love to work with—exists at one, and only one, very specific frequency, and handling this mathematically requires an abomination known as the Dirac delta function.

But I'm getting ahead of myself. We can forget about Dirac's mathematical monster, fascinating though it is, until much later in the book.

More to the point, then, why should Fourier analysis, this apparently innocuous shift from the time domain to that of frequency, be at the heart of so many areas of physics? And how does it connect Heisenberg with Helloween, *Hell Awaits*, and *Hell Hole*?

Thus far, we've only considered the frequency spectrum of single whistled notes. We could, of course, whistle two different notes together at the same time (to form a harmony) and, as you might guess, we'll see two peaks in the frequency spectrum . . .

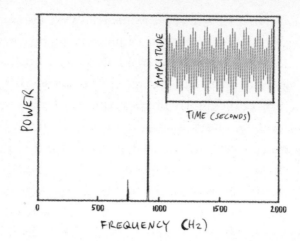

The inset shows how the harmony whistle waveform varies in time. It's rather more complicated and difficult to interpret than the simple two-spike frequency spectrum. Given just the waveform without its frequency spectrum, you'd be hard-pressed to tell that it was made up of just two sine waves mixed together. This is the beauty of Fourier analysis—it helps us deal with complexity by representing a signal in a much more palatable and easier-to-digest form.

It's time we increased the metal quotient just a little, however, starting with revisiting the guitar note that opens "Sanitarium." It's just a single note, so just a single spike in the frequency spectrum, right?

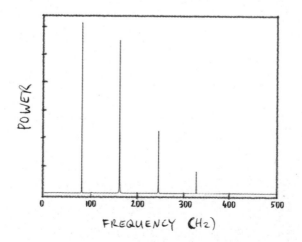

Wrong.

The frequency spectrum of that opening E in "Sanitarium" has multiple spikes. That means that there's more than one frequency present. But isn't it a single E note? Yes, it is indeed a single E note. Remember the sheet music from Chapter 1—an E on guitar is represented just like an E on piano. It's the same note.

So what's going on? Why so many distinct frequencies for a single note? To understand the origin of the multiple spikes in this frequency spectrum, it is time to consider resonance in rather more depth.

Respond, Vibrate, Feed Back . . . and Resonate[11]

If there's a band that really resonates with me, it's Rush. There are many reasons for this—their musicianship, their arrangements, their passion and emotion, Neil Peart's lyrics (although, it has to be said, he has produced a few stinkers among the classics[12]), and their understated intelligence and humor. (Oh, and that they thought it was a good idea to write a song called "By-Tor and the Snow Dog." That alone means they'll always be heroes to me. "*Tobes of Hades, lit by flickering torchlight. The netherworld is gathered in the glare.*" Pioneering and classic prog metal. I still get goose bumps.)

But another reason Rush made an especially strong impression on me was that they embedded sci-fi concepts in their music. As a child of the '70s and a teenager of the '80s, I was, of course, a major fan of *Star Wars* and *Star Trek*, but I had (and still have) a particularly strong affinity for British sci-fi (wobbly sets and wooden acting notwithstanding). I adored *Blake's 7*—a dark and chronically underfunded sci-fi/

---

[11] From Rush. "Chain Lightning." *Presto.* Atlantic Records, 1989.

[12] Stinker: *"Net boy, net girl. Send your signal 'round the world."* Classic: *"Somewhere out of a memory of lighted streets on quiet nights . . ."*

space opera show about a bunch of anti-heroes battling the evils of the totalitarian Terran Federation[13]—and watched it fanatically twice a week.[14] Alongside Blake et al.'s shenanigans with the Federation's supreme commander, Servalan,[15] I tuned into *Sapphire and Steel*, *The Hitchhiker's Guide to the Galaxy*, *Doctor Who* (with Tom Baker), and devoured issues of *2000 AD* (which featured such iconic characters as Judge Dredd, Dan Dare, Rogue Trooper, and Sláine).

So when I discovered Rush's *2112*—a concept album based on the premise of a dystopian future where another totalitarian regime, this time the Priests of the Temples of Syrinx,[16] have *"taken care of everything. The words you read, the songs you sing. The pictures that give pleasure to your eyes"*—I was hooked.[17] The hero of the piece discovers an old guitar (in a room beneath a cave behind a waterfall) and is spellbound:

What can this strange device be?
When I touch it, it gives forth a sound
It's got wires that vibrate and give music
What can this thing be that I found?

(S)he brings it to the Priests, but they're not impressed—the guitar should be forgotten, a "silly whim" that's "another toy that helped destroy the elder race."[18]

---

[13] *Blake's 7* was created by Terry Nation, whose other major contribution to the canon of British sci-fi is Doctor Who's arch-nemesis—or should that be arch-nemeses?—the Daleks.

[14] We didn't have a video recorder at the time but, fortunately, the national Irish broadcaster RTE (Raidió Teilifís Éireann) also transmitted *Blake's 7*—on a Sunday (if I recall correctly), whereas the BBC episode was on a Monday or a Tuesday—so I could watch it biweekly.

[15] She *certainly* made an impression.

[16] Well, technically, half a concept album. "2112," the song, took up side one of *2112*, the album.

[17] Without giving away too many spoilers, in his novel *Ready Player One*, Ernest Cline completes the sci-fi–Rush circle, making the "Holy Trinity" of Lee, Lifeson, and Peart a key part of the plot.

[18] Peart doesn't reveal the gender of the central character at any time in the lyrics.

The middle section of *2112*, a twenty-minute-long prog metal classic, is entitled "Discovery" and it begins with the sounds of our hero striking guitar strings at random and then tuning up. We can hear that considerable use is being made of harmonics to tune the strings—our hero was clearly a prodigious and natural talent to have grasped the importance of harmonics so early in his or her exposure to the guitar. (We'll be covering harmonics in depth very soon—stay tuned.)

When I was first learning to play, I loved the "Discovery" section of *2112* because I could pick along with the first few bars despite knowing virtually nothing about guitar. And, just as at the start of "Sanitarium," an open E string is sounded in "Discovery" (when the guitar is being tuned—the piece then shifts to the key of D-major). Many guitarists love the sound of open strings like those played at the start of "Discovery" and "Sanitarium" because of the ringing and resonant "flavor" of the notes. Alex Lifeson, in particular, has made clever and compelling use of open-string–based chord voicings *and* harmonics throughout his career in Rush. There's even an "archetypal" voicing known among Rush aficionados as the "Alex chord" (or the "*Hemispheres* chord") because it's so representative of his approach.[19]

As you might suspect, there's some fascinating physics behind the resonant tones produced by Lifeson (and, of course, this extends to the notes and chords generated by other, somewhat less godlike, guitarists). What you might not have suspected, however, is that there are deep connections between those musical notes and just how particles behave when they are constrained at the quantum level. Undergraduates in their first quantum course, be it in physics, chemistry,[20] or any other discipline, spend a great deal of time on the striking parallels between

---

[19] For the musicians among you, it's an *F♯7 add 11* chord. This is basically a standard F♯ major chord but with the top two guitar strings, B and E, not fretted as usual and instead left open. (For a standard F♯ major chord, the E and the B are fretted to sound an F♯ and a C♯ note, respectively.) It's a fantastically evocative chord for Rush fans as it's the opening to *Hemispheres*.

[20] All of chemistry is just quantum physics at heart. (Joke. *No really, Prof. Amine, it's a joke. I promise. Please put the ring stand down!*)

standing waves on a string and the properties of electrons confined to a region of space. This confinement could be that within an atom of some garden-variety element off the periodic table (hydrogen [H], helium [He], neon [Ne], gold [Au], lead [Pb], etc. . . .[21]) or, rather more controllably and excitingly, that of an electron that has been trapped in an *artificial atom* (aka, a quantum dot).

We'll get back to artificial atoms in due course, but to uncover the "hidden analogies," as Fourier put it, between guitar notes and quantum confinement, we first have to dive just a little deeper into wave motion to understand the physics of guitar notes, be they Lifeson's shimmering suspended chords or the guttural grinding of Meshuggah's seven strings. Only then will we begin to see why that opening note of "Sanitarium" contains so much more than meets the ~~eye~~ ear.

Echo Beach

Having covered the distinction between longitudinal and transverse waves in Chapter 1, and also having thought a little about kinetic and potential energy, we now need to consider one other key characteristic of wave motion: traveling vs. standing waves. When you're standing in the middle of a crowd at a festival, being simultaneously drenched by a downpour of rain or sleet or hail or snow or mud or urine (or any combination thereof) and blasted by the 137 decibels generated by the band onstage,[22] the sound energy is transferred to you via a *traveling* wave. "Traveling" means exactly what you might think—the wave travels from its source to its receptor (in this case, your ears) carrying sound energy with it, a bit like throwing a stone in a lake generates a disturbance that

---

[21] But not sapphire. Or steel. (That's a *very* nerdy in-joke for fans of that aforementioned classic sci-fi series featuring David McCallum and Joanna Lumley in the title roles.)

[22] The decibel is a measure of sound intensity. We'll discuss just how it's defined later on.

freely propagates outward, shaking up the water as it moves.[23] But this is rather different from what happens on a guitar string. There, the wave generated by plucking the string is *confined* between the nut and the bridge of the guitar: it's got nowhere to go.

So if the wave is trapped and can't freely move like its traveling counterpart, what happens? We're going to reflect on that question by imagining a wave encountering a solid obstacle.

Let's do the following *gedankenexperiment*.[24] Find a beach next to a cliff. Set up your guitar amplifier about 100 meters from the base of the cliff. Don't worry about (a) whether or not you're a guitarist, (b) whether you have a guitar and amplifier on hand, and/or (c) where the heck you're going to find a power outlet on the beach. This is a *gedankenexperiment*, after all. Switch on and wind up the volume to the maximum.

---

[23] The analogy doesn't quite work because waves traveling on the surface of the lake aren't longitudinal, they're transverse. But let's not be too literal minded just yet.

[24] The *gedankenexperiment* (direct translation: thought experiment) has a long and distinguished history in quantum physics (and, indeed, in physics in general). Long before we could image and manipulate single atoms and molecules, the quantum pioneers (Dirac, Schrödinger, Einstein . . .) spent a great deal of time debating the potential outcomes of sophisticated experiments, impossible at the time, on these particles. The development of relativity was also driven by a variety of imaginative and creative *gedankenexperiments*.

Now play that wonderfully crunchy chord at the start of the signature riff of AC/DC's "Back in Black."

What will happen to the sound wave generated by that guitar "crunch"? It'll travel outward from the amplifier, hit the cliff face, and be reflected—you'll hear this reflection as an echo. We can easily work out when you'll hear the echo. Sound in air travels at about 1,225 km/h (or, for those of you who prefer nonscientific units, about 770 mph). This is ~340 meters per second. Given that you and the amplifier are 100 meters from the cliff, the sound has to travel 200 meters (there and back). Speed is just the distance traveled divided by the time taken, so if we do the calculation, we find that you'll hear the echo a little over half a second after you hit the chord.

We can express all of that much more compactly using a smattering of algebra, of course. Let's call the speed of the wave v, as is traditional in physics (to denote velocity).[25] The distance will be denoted by d and the time taken for the wave to travel by t. So,

$$v = \frac{d}{t}$$

Rearranging that formula we have:

$$t = \frac{d}{v}$$

Plug the numbers in (d = 200 m, v = 340 m/s) and, bonzo, we find that t is 0.588 seconds (or 588 milliseconds).

Now wheel the Marshall stack closer to the cliff face. The time at which you hear the echo will decrease directly in line with the reduced distance the wave has to travel. This is another example of the type of linear relationship that was mentioned in the context of our drum beater's motion last chapter. If we were to measure the echo time for different Marshall-to-cliff distances, and then plot a graph of those measurements,

---

[25] When the speed we are talking about is the speed of light, however, it gets its own special variable, c. The reasons behind the choice of c are lost in the mists of time but it may be related to the Latin word for speed—*celeritās*. Or, more prosaically, c might simply be shorthand for constant.

we'd get a straight line. And from the gradient—steepness—of that line we could work out the speed of sound.[26]

The cliff face represents just one reflecting surface. It gets rather more complicated if we set up a situation where sound bounces off a number of surfaces, leading to many more echoes and a plethora of waves. (Depending on your perspective, this is either a fascinating phenomenon in acoustic physics or a massive pain in the, errmm, ears; sound engineers and those involved in audio recording in general tend to fall into the latter category because dealing with wave reflections can be the bane of their lives.) When waves meet each other, they combine and *interfere*. If the peaks and troughs of the two waves line up, as sketched below, this is called *constructive* interference; when the peaks of one wave align with the troughs of another, we have *destructive* interference—the waves cancel each other out.[27] The extent to which the peaks and troughs of the waves line up is captured by a value called the *phase* of the wave: with constructive interference the waves are entirely in phase; destructive interference means that the waves are fully out of phase. Between these two extremes, the phase difference between the waves establishes the shape of the final wave.

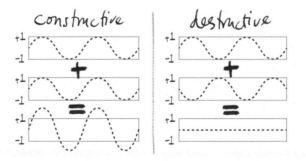

[26] As we continue to decrease the distance between the amp and the cliff face, there'll come a point at which the echo time is so short we won't be able to distinguish the echo from the original signal. This is a fascinating psychoacoustic phenomenon, known as the precedence effect, which was first studied by the German scientist Hans Wallach and his colleagues.

[27] You might ask just where the energy goes if two waves destructively interfere. That's a great question—lots of physicists ponder it as students. It's addressed in the appendix.

Sin After Sin[28]

We can better understand phase by returning both to the drum beater and to Marsha the Mosher, whom we met in Chapter 2. I said at the time that the equation we use to capture the motion of a wave goes like so:

$$\theta(t) = A \sin(\omega t)$$

Well, I fibbed. Just a little. Sorry. There's one term left out of that equation that we need to introduce in order to accurately capture the effects of constructive and destructive interference: phase. Phase is essentially a method of keeping track of the starting position (or starting time) of a wave. In Fourier analysis we can often get the essence of the physics without being too concerned about phase but, eventually, to understand the quantum world, we can't avoid phase.

In keeping with the usual traditions in physics and maths, we're going to use the Greek symbol φ (pronounced "phi") to represent phase.[29] Here's how we incorporate phase in the equation:

$$\theta(t) = A \sin(\omega t + \phi)$$

Note that φ is included in the "input" to the sine function. It's an angle—indeed, "phase" is shorthand for "phase angle." The phase angle is measured in radians, so let's ask our mosher to start dashing around the circle pit again from different starting positions, and see what effect different values of the starting position (relative to the x-axis this time), φ, have on the sin(e) function. Sin after sin, if you will . . .

---

[28] A classic Judas Priest album for many reasons, not least the groundbreaking metal vocals and pummeling double bass drums of "Dissident Aggressor" (which Slayer later went on to cover in their own inimitable style).

[29] Phi also represents the golden ratio (although generally it's the uppercase Φ that's used)—an irrational number (1.61803398 . . .) that crops up in a wide range of fields spanning architecture, engineering, mathematics, music, and art (even when it shouldn't). More on that type of math metal numerology in a later chapter.

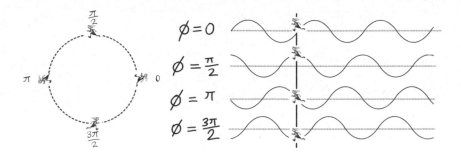

From the sketch above we can see that changing the phase angle shifts the sine function along the x-axis. Now, the x-axis could represent time, as has been the case for the drum beater and the various musical notes we've looked at thus far. *Or* it could represent position. In other words, we can look at how a wave changes in time at a specific position in space—measuring the variation in the volume of the gig at seat number C5 in Metallizer's arena show, for example—or the sine function can also represent how the wave varies in space at a particular instant of time. For the latter, what we have is a snapshot of the wave in space.

In other words, changing the phase just means that the wave shifts position: the peaks and troughs move forward or backward in time and/or space. Now, while this might encouragingly sound rather like some method of invoking time travel, it's unfortunately not. I should be very clear by what I mean by a wave shifting in time.

Here, for absolutely the last and final time, is a graph of that first whistled note we considered. (Yes, I know I'm trying your patience, but please bear with the whistling for just a little longer.) The graph at the top shows that the wave is zero—no volume—at the 100 millisecond mark (shown right in the crosshairs). But in the bottom graph, the wave has its *maximum* amplitude at 100 milliseconds. We say that one wave is phase shifted with respect to the other. In this case the phase difference is $\pi/2$ radians (or 90 degrees in old money).

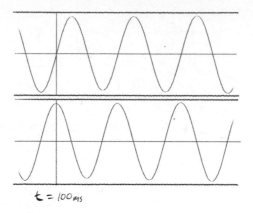

$t = 100\,ms$

That's all there is to phase. Really. It's just a way of measuring how much two (or more) waves are in or out of step with each other. But the simplicity of the concept belies the central importance of phase in just about everything in our universe, including the dynamics of every guitar note that's ever been played.

Chapter 4

RUNNING TO STAND STILL[1]

Patterns emerge in nature's dance
Numbers are born in the wheel of chance
Why do I see this?
What does it mean to me?
A grand design in all its majesty
Vibrating strings . . .

—from Ayreon's "The Prodigy's World"[2]

---

[1] Some of you will recognize this chapter title as a U2 song title (from *The Joshua Tree*) and may now question my commitment to the metal cause. I fully realize that for many, U2 is the antithesis of metal music—a metal antichrist, if you will. But, that song title is so apt for this section of the book that I really couldn't resist using it. And, in any case, I have a soft spot for the first five U2 albums. (I know, I know. I'll hand in my patch-laden denim jacket on the way out . . .)

[2] *The Theory of Everything* (2013).

Finally—finally!—we have all of the pieces of the puzzle in place, everything we need to understand what happens when a guitar string is plucked, struck, thumped, caressed, or otherwise cajoled into producing a note. Unlike the traveling sound waves we've discussed thus far, the wave on a guitar string isn't going anywhere: it's trapped between a rock and a hard place (well, between the bridge and the guitar nut). There's a solid obstacle at each end of the string, meaning reflections play a key role in the dynamics of string-bound waves.

Let's reflect on precisely what happens when a wave is reflected. A sound wave smashing into the wall of a concert hall is reflected in a very different way than a wave on a guitar string that encounters the nut at the end of the neck. When Rob Halford hits one of his patented metal screams and it spreads out through an arena, there's no change in the phase of the wave carrying the Metal God's shriek when it hits a wall (or any other solid obstacle). A higher density region of the wave is reflected as . . . a higher density region.

But that's not what happens when a wave traveling on a guitar string encounters the nut at the end of the neck. In that case the phase of the wave changes by π radians: the waveform is flipped upside down. (Look back at the "sin after sin" graphic on page 55.) This reversal doesn't just happen for waves on guitar strings—a light wave, a radio wave, or, indeed, any type of electromagnetic or electrical signal will be phase shifted like this when it reflects under the appropriate circumstances.[3] Luckily, we don't need to consider a complicated example like an electromagnetic wave to see this phase shift in action: applying a thwack to our vintage heavy metal Slinky—*"The hit of the day when you're ready to play . . ."*—is enough to see the phase shift in action.

The cartoons below show what happens when a Slinky is given a thwack to generate a pulse that subsequently hits a solid obstacle—in this case a wall it's attached to—and is reflected. Note that the pulse flips around when it hits the wall. We just have a single pulse of energy on the Slinky, rather than a full-blown wave, but the same principle applies for a complete wave on a string: it's turned on its head when it meets a rigid obstacle. Why does this happen? (And does it always happen when a wave on a string is reflected?)[4]

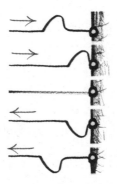

---

[3] Those "appropriate circumstances" are subtle and multifaceted and are the subject of entire books (and series of books) on their own.

[4] Some of you might quite understandably be a little confused about reflection at this point. Didn't I say that there was no phase change when the sound wave from Rob Halford's shriek hit the wall? (Yes, I did.) Yet now I say that there is a phase change for a wave on a string! Is this not inconsistent? Well, no. The fact that the string is fixed at the end is what gives rise to the phase change—it's all about the boundary conditions. If the string was free to move at its end (i.e., transverse to the wave direction) then there wouldn't be a phase change.

What we're seeing when the Slinky pulse inverts is Newton's third law in action: "For every action, there's an equal and opposite reaction."

In the case of the Slinky, the wall applies a reaction force that opposes the force of the wave. The net result is that the mechanical pulse flips over—or, in more technical language, its phase is shifted by π radians—when it's reflected. The same holds true for a train of pulses, one after another, or a wave on the Slinky: reflection produces a phase shift of the wave. And it's exactly the same physics in action on a guitar string—a pulse or a wave on the string will have its phase shifted by π radians when it encounters the nut or the bridge at the ends of its travel.[5]

NEWTON'S LAW OF ACTION-REACTION

"Reaction" has a particular meaning in physics that is much more specific and precise than its interpretation in non-scientific, everyday language.[6] You may be sitting in a chair as you're reading this book (or perched atop a barstool, or reclining on a beanbag, or whiling away some time between bands at an outdoor rock festival as rivers of mud flow by). The force of gravity is acting on you, yet you've

---

[5] For other types of waves and other types of barriers/obstacles, this phase shift doesn't occur. But we don't need to worry about those just yet in the context of the metal-physics interface.

[6] Physics is awash with examples of scientists co-opting language like this. It can be confusing at times. A great example is the difference between the physics definition and everyday usage of "work." Pick up a book and hold it a meter or so above the ground. Raise it above your head. Now bring it back to the starting position. Do that ten times. In the context of textbook physics, *you've done no work*. To a working physicist, work doesn't bring to mind the grind of the nine to five but is, instead, the path-dependent integral of the scalar product of the force and displacement vectors. (We're usually fully capable, however, of understanding the context in which the word "work" is being used. Integrals are generally not the first thing to spring to mind when a friend tells us that they've got no time for living because they're working all the time.)

not fallen through the floor and sped off toward the center of the Earth due to the gravitational attraction. Why is this?

The entirely obvious answer, of course, is that the floor gets in the way. The slightly less obvious, but more physics-y, answer is that the floor (or chair, or stool, or mud-encrusted ground) is applying a reaction force to you that acts in opposition to the force of gravity: same magnitude, different direction. Net result: you don't go anywhere. (The fundamental origin of this reaction force is fascinating. Ultimately, it arises from the antisocial nature of electrons, otherwise known as the Pauli exclusion principle. Electrons do not want to be squeezed into the same state—the electrons in the floor and those in your body repel each other very strongly. And not just because they're all negatively charged! The Pauli exclusion principle is quantum statistics in action—but that's a whole other story . . .)

At this point you might reasonably think that when a guitar string is plucked, the resulting motion will be a complex mess of waves bouncing back and forth and running into each other. Isn't that going to produce exceptionally complicated waveforms? It turns out that the answer, rather surprisingly, is no. Nature is kind to us—out of all of that complexity comes a breathtakingly simple result. Those waves that travel up and down the string reflect and interfere, constructively and destructively, and the net effect is not a mess at all, but instead a rather plain and unfussy type of wave, a wave that stands still: a *standing wave*. (Let it not be said that we physicists lack imagination when it comes to thinking up names for different phenomena.)

The interference of the original and reflected waves on the string results in oscillations whose peaks and troughs always stay at the same position. That's why the U2 song title that heads up this chapter is especially apt: the waves run to stand still. As we've seen above, because phase determines how the peaks and troughs of waves line up—and, thus, how they interfere—the phase shift of π radians at the ends of the string is exceptionally important in the formation of standing waves.

What does a standing wave look like? Here's how we traditionally sketch them . . .[7]

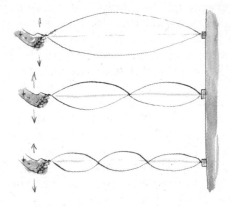

And here it is in a not-so-traditional setting . . .

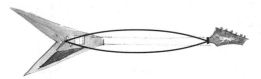

When the string is plucked, the oscillation will decay off, just as the drum beater gradually came to a halt in Chapter 2. But the peak of the oscillation will always be found at the center of the string (in this case). This is why it's called a standing wave—it's as if the wave were fixed in space. Although the wave is standing on the string, however, it nonetheless produces a traveling wave in the air around it. Even without plugging in the guitar, and thus converting the standing wave to an

---

[7] If you're not a physicist, you may remember this sketch from high school physics lessons. If you *are* a physicist, that standing wave diagram needs no introduction—it's an old and exceptionally familiar friend that appears under many different circumstances.

electrical signal that can subsequently be amplified, the sound arising from the vibrating string can be heard; the molecules in the surrounding air are displaced back and forth. The transverse standing wave on the string produces a longitudinal traveling wave in the air.

When we plug in, it's a rather more involved process: the standing wave on the string generates an electrical signal via the pickups of the guitar that is usually fed through an effects pedal (or a bank of effects pedals/digital signal processor) into an amplifier. The amplifier in turn drives a loudspeaker, which moves the air molecules back and forth to a much greater extent than the weedy string could ever do by itself. The net effect is again that the transverse standing wave on the string is converted to a longitudinal traveling sound wave that bangs against the eardrum (and which in turn may result in banging of the head).

Setting Boundaries

If this were a classroom lecture, at this point I'd write down—or better, derive—the equation that describes the motion of the string.[8] But we don't need to dig that deep into the mathematics to understand the underlying physics.[9] The standing wave that we've considered is just *one* standing wave solution to the equation that describes the motion of the string. There are many, many different solutions. (In fact, mathematically, there are an infinite number. Fortunately, physics reins in the maths.)

What all those solutions have in common is that the string is fixed at the ends: it's held in place at the bridge and at the nut. Our *boundary conditions* are that the string doesn't move at those positions or, in other words, the wave has a value of zero there. This means that we have a *family* of standing waves that can form on the string, and the trait that's

---

[8] The wave equation is explained in the "Maths of Metal" appendix.
[9] Mathematics is the language of physics. But physics isn't mathematics (and vice versa).

common to all of them is that they have a node at either end. Let's sketch a few of those standing waves. You might recognize the shapes . . .

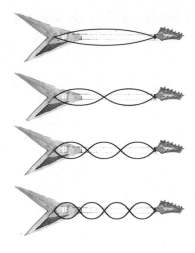

It is indeed our old friend the sine wave. Again. We call each of those sine waves a mode of vibration of the string or, more succinctly, a *harmonic*. But when we pluck a string we don't just get one mode of vibration—one harmonic. Why would we? The maths tells us that any of the harmonics sketched above (and all the others we haven't included) would be a valid solution to the equation that describes the motion of the string. Now, we could get a computer to solve that equation (or we could attempt to do it ourselves with pen and paper). But the string itself *is* a computer of a sort—it's a physical computer that solves the equation for us. And the solutions are as sketched above.

When we pluck the string, it vibrates in lots of those modes *at the same time*. But note that the only standing waves that can form are those that meet the boundary conditions—there has to be a node at each end of the string. Only waves that line up neatly so that they have an amplitude of zero at our two nodes are possible. And that means that only certain wavelengths are possible. In other words, the wavelengths are *quantized*. That's where the quantum in "quantum physics" comes from: instead of having any old value of a quantity, we can only have specific values of that quantity. Other values are verboten.

Very often, "quantum" is used in the context of weird, wacky, or netherworldly phenomena that have no counterpart in the everyday world around us. But here we see that the waves on the humble strings of, say, the less-than-humble Yngwie Malmsteen's strat are nothing less than quantization in action. No matter how accomplished a guitarist Mr. Malmsteen might be—and he's inarguably about as accomplished as it gets when it comes to neoclassical neck noodling—he can't beat physics; only the waves that match the boundary conditions can form on the strings he's playing. Every lick and riff he plays is constrained by the node created where he frets a note and, at the other end, by the bridge of the guitar. He can't beat those boundary conditions. Luckily, despite these "limitations," there are very, very many wavelengths Malmsteen, and all other guitarists, can select to make their music.

We need only a smattering of maths—even less than what we considered when we were thinking about the sine wave equation—to work out what wavelengths are possible for standing waves on a string. (We'll stick with the open strings, but the same principle holds for fretted notes.)

Let's say the string has a length L. We've already seen what the first mode of vibration of the string, the very longest possible wavelength—the fundamental—looks like. It's this:

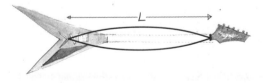

In other words, we have a node at each end, and an *antinode*—a position where the amplitude of the vibration reaches its maximum—in the middle of the string. But we only have half a cycle of the full sine wave on the string in this case. So, if the string has a length L (as shown above), the wavelength for this fundamental mode must be $2L$. (If L is only half a wave, it follows that the whole wave would be twice that.) Opinion is divided—largely between physicists and engineers—as to whether we call this fundamental mode the first harmonic (for reasons that will become clear in a second). Whether opinion is divided or not, *I'm* certainly going to call it the first harmonic (because it's my book, and engineers refer to the imaginary number i as j and so are clearly not to be trusted in matters of nomenclature[10]).

Here's what the second harmonic (the second mode of vibration) looks like:

And the wavelength this time is just L.
One more. Here's the third mode:

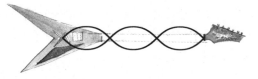

---

[10] To be fair to engineers, particularly those of the electrical/electronic persuasion, they have to deal with currents quite a bit, so they use i to represent the flow of electricity. That meant they needed to find another symbol to represent the square root of -1, usually written as $\sqrt{-1}$, so they chose j. The only problem with this is that physicists (and, indeed, quite a few engineers) also tend to use j as a symbol for current density (the amount of current following through a certain cross-sectional area [for example, 1 mm^2] of a component/object). *Sigh.* These conflicts in terminology/nomenclature between different areas of science and engineering are frustratingly confusing for anyone trying to learn the subject(s). When you're a little longer in the tooth, you just accept it as a "cultural" difference, much like that "tomato, toe-mah-toe" thing. (And don't get me started on the differences between conventions in thermodynamics among the physics, chemistry, and engineering "tribes.")

The wavelength this time is ⅔ the length of the string. We can write that in a slightly more compact way like this:

$$\frac{2L}{3}$$

Now, we could continue doing this harmonic by harmonic, each time sketching the mode of vibration and working out the value of the wavelength, or we could look for a general equation that relates the harmonic number (first, second, third, and so on . . .) to the wavelength for that harmonic. Guess which approach we're going to adopt?

Physicists are very fond of stating bluntly that something can be seen "by inspection"—that is, there's no real mathematical work or detailed derivation required; the answer becomes obvious after just looking at the equations and/or graphs for a while. Perhaps a useful analogy is with solving a clue for a crossword puzzle. Many undergraduate physics students, however, get very irritated indeed when a professor casually drops the "by inspection" bombshell into a lecture. Usually it means that when they later go back over their lecture notes, they'll spend an hour working out just how they were meant to grasp that "obvious" piece of maths or physics simply by inspection. This is because what's obvious by inspection to someone who's taught physics for twenty/thirty/umpity years can be rather less obvious to someone learning the subject for the first time.

But in this case, I'm afraid that I am also going to advise the trusty old "by inspection" strategy. That's because there's no better way of understanding the relationship between the wavelengths of those standing waves than to make sure you can see where the following equation comes from. Take a look at the sketches of the shapes of the harmonics on the previous pages (and/or do some sketches yourself) and convince yourself that the following equation is correct. By inspection.

$$\lambda_n = \frac{2L}{n}$$

In that equation, L, as we saw above, is the length of the string. λ (the Greek letter lambda) represents the wavelength of a standing wave, and λ_n stands for the wavelength of the nth harmonic. The n is what we use in maths to mean "insert number here," in this case, the harmonic number; $n = 1$ is the first harmonic, $n = 2$ is the second harmonic, etc.

Let's try plugging in a value of n as a test. λ_1 is the wavelength of the fundamental, or first harmonic. Plug in 1 for n on the right-hand side of the equation and what do we get? Well, we find that $\lambda_n = 2L$, exactly as we saw from the sketch of the fundamental mode. Try plugging in a few more values for n and you'll see that the simple equation above is all you need to determine the wavelength of any harmonic.

But remember that what we're really interested in is frequency. How do we translate the wavelength into the frequency for that particular harmonic? After all, it's the *rate* at which that string moves back and forth that will determine the pitch of the note we hear. Musically, we don't really care that much about wavelength on the string, although it's at the very core of the physics of how a note on that string is produced. We want to know the frequency . . . Kenneth.[11]

Fortunately, there's a very simple equation relating wavelength and frequency:

$$\mathrm{v} = f\lambda$$

In that equation, v is the speed of the wave, f is the frequency, and λ represents, as we know, the wavelength.

The speed of the wave on the string, v, is determined by its tension and mass per unit length (e.g., how many grams are there for 1 cm of the string?). For the moment we'll assume that those quantities don't change. It's generally the case that the tension on a string remains constant while a note is being played, but that's not true for a lot of metal guitar solos. Kerry King, for one, has made a career out of wildly varying the tension in his strings when he's soloing. His whammy bar exploits are core to the sound of Slayer and involve rapidly and regularly varying the tension in

---

[11] Given that I've already sold my soul and included a U2 reference, I'm hoping that an R.E.M. allusion will now easily pass under the radar.

the strings, continuously varying the pitch of a note. There's no better example of this than the solos in "Angel of Death."

But, for now, let's assume that we're not whacking the whammy like Mr. King. Nor are we changing to a different type of string mid-song. This means that the tension and mass per unit length are fixed. And that, in turn, means that the wave speed on the string is constant. In fact, we could work out the speed of the wave from knowing some parameters about the string. But we'd really prefer not to have to take the strings off the guitar and weigh them to work out the mass per unit length. And measuring the tension in the string is also going to be a bit of a faff. So let's work out the speed of the wave on the string from properties we can measure much more easily.

Here's the Fourier spectrum of the opening note of "Sanitarium" again, but this time the spike related to the fundamental (the lowest frequency harmonic) has been labeled as f_1. (The other harmonics in the spectrum are labeled f_2, f_3, and f_4. We'll get back to those in a moment.)

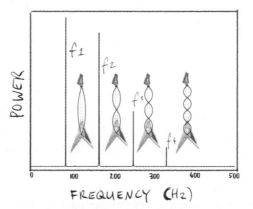

So now we know the frequency *and* the wavelength of the fundamental. (Remember the wavelength of the first harmonic is $2L$.) This means that without having to do any mucking about measuring the tension or mass per unit length of the string, we can work out how fast waves move on that string:

$$v = f_1 \lambda_1$$

We know the value for f_1 because we can read it directly off the graph on the previous page—it's about 80 Hz. If we were to zoom in on the spectrum a little more, and measure slightly more accurately, we'd find that it's actually a frequency of 82 Hz. To work out the value for λ_1 we first need to know the distance from the bridge to the nut on the guitar. In this case it was a Les Paul copy, so the scale length—as the nut-bridge distance is called in the trade—is roughly 24.75".[12] We need to convert that to standard scientific units, which is 0.62 meters. This is the value of L but you'll remember that $\lambda_1 = 2L$. So the wavelength for the first harmonic (the fundamental) is 2×0.62 m = 1.24 m. So, the speed of waves on the string is 82 Hz × 1.24 m/s or roughly 102 m/s.

Even with just the two key equations we have thus far, we can deduce much more than just the wave speed. A key result that falls out of the maths is that the frequency of the second harmonic is twice that of the fundamental, the frequency of the third harmonic is three times the frequency of the fundamental, and so on. As ever, we should check that the predictions of our mathematical theory hold up. And, indeed, the spectrum on the previous page verifies the maths. We can measure the frequencies directly from the positions of the spikes on the x-axis. The fundamental has a frequency (f_1) of about 80 Hz, the second harmonic (with frequency f_2) is at roughly 160 Hz, the third harmonic is at roughly 240 Hz, etc. . . .[13] In other words, the frequency of the nth harmonic is n times the frequency of the fundamental. This very simple relationship

---

[12] Even in the UK and Ireland, it's common to refer to the length of guitar necks in that archaic Imperial system of units so beloved by our American cousins.

[13] I'm using the word "roughly" very deliberately here. There's a slight complication (that we don't really need to worry about too much, but it's covered in the appendix for those interested in the real nitty-gritty of the physics). Strictly speaking, the fact that the nth harmonic is n times the frequency of the first harmonic holds only for an idealized perfect string. That type of string exists in the mathematical universe but not in the real world. A real string, like that on the guitar used to record that opening note of "Sanitarium," is not perfectly "floppy"—there's a stiffness, and this adds a small correction to the value of the frequency. This is a minor consideration in terms of the central physics principles

between the frequencies of the various harmonics underpins the "musicality" of a guitar (and any other stringed instrument): if the harmonics weren't integer multiples of each other, the guitar would be less of a musical instrument and more of a noise generator. And although some would claim that metal is nothing but noise, we know better. (Kerry King's guitar solos notwithstanding.) Also included in the spectrum on page 69 is a sketch of the shape of the various standing waves. Note that each harmonic has a different number of nodes and antinodes. Compare the fundamental with the second harmonic: for the latter, there's a node in the middle of the string. We can force the guitar string to resonate so that this harmonic is accentuated by ensuring that there's a node at just the right place. And we can make sure there's a node where we need it by lightly resting our finger on the string at the correct position. For the second harmonic, this is above the twelfth fret—bang in the center of the string.

🔊 **If you compare the sound of the note generated this way to that we get when plucking the open string (here: https://www.youtube.com/watch?v=rHpYFtDmW6E), you'll notice the tone of the latter loses some of the character of the guitar sound.**

And if we compare the frequency spectra for the two notes, we find a dramatic difference:

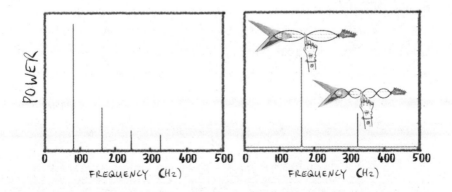

here, but I'd be remiss not to point it out. (We'll see, particularly in Chapter 6, that this "layers-of-approximation" approach to science is very common.)

By lightly touching the string at its center, we stop it moving there—we enforce a node. That means that only those harmonics that have a node at the center of the string can exist on the string. And that in turn is why the spikes for the first and third harmonics are absent in the spectrum on the right. Only the second and fourth harmonics are present in the spectrum because only the second and fourth harmonics[14] have a node in the center of the string. By ensuring that there's no vibration at the center of the string, we've "filtered" out the standing waves that don't have a node there.

This approach to generating what are called natural harmonics is exploited throughout metal and rock. Rush's "Red Barchetta" opens with a riff based entirely on harmonics played on the G, D, and A strings. If we compare the waveform for the opening natural harmonic (at the twelfth fret on the G string) to the signal for the open G string, we again find that the latter is a very much more complex signal with many more harmonics, as shown on the next page. On the left we have the spectrum of frequencies for a G string that is just plucked. The inset shows what happens when we do the same thing as before and touch the string lightly above the twelfth fret (in the middle of the string)—we filter out harmonics that don't have a node there. The sound of the string becomes much more "bell-like" in tone—it is much less harmonically rich and therefore it's becoming closer to a pure sine wave in character. By filtering out those harmonics that give character to the note, the guitar loses some of its identity. That's why the waveform for the opening note in "Red Barchetta" looks much less complex than when we simply pluck the open string, as shown in the two graphs on the right on the next page.

---

[14] . . . and sixth, and eighth, and tenth, and so on. But those harmonics aren't visible in the spectrum in any case—the frequency range isn't wide enough, and the way I've chosen to plot the data (on a linear rather than a logarithmic scale) masks those contributions to the note.

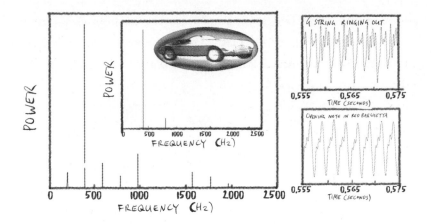

Again, this is Fourier analysis in action. The open string vibrates in a variety of different modes; the resulting waveform is the sum of the shapes of all those harmonics. Think of Fourier's approach as painting with sound: sine waves of different amplitude, frequency, and phase make up the palette, and any pattern can be created by mixing the audio colors in just the right way. By summing up different sine waves, complex patterns of sound are created; and by removing sine waves, we simplify the signals.

The waveform for the open string is the sum of the shapes of all the harmonics that are physically possible for the unfretted G. By forcing a node to be present in the middle of the string—as Alex Lifeson does for that opening natural harmonic in "Red Barchetta"—there are fewer harmonics, i.e., fewer sine waves, contributing to the note, making it necessarily less complex.

Don't Fear (the) Fourier

Let's now push the dials to eleven and beyond, and look at harmonics in a very metal context: Fear Factory's "Scumgrief." This is a track from their debut, *Soul of a New Machine,* which features a riff where natural harmonics play an integral role. (For those of you with the song at

hand—or if you can find it on YouTube—the harmonics appear for
the first time at 0:58 into the track.) Here's the waveform and the frequency spectrum for those harmonics, first played on a guitar with no
distortion:

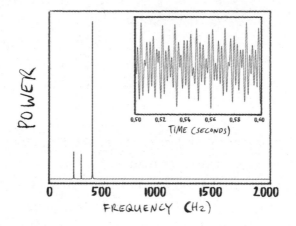

There are three clear peaks in the Fourier spectrum (at approxi-
mately 215 Hz, 284 Hz, and 388 Hz) that, when added together in just
the right way, produce the "Scumgrief" waveform shown in the inset
above.[15] When I say "just the right way," I mean that not only do we
have to use the correct "amount" of each sine wave (which we can work
out from the height of the spikes in the spectrum), but we need to know
the phases of the waves. In other words, we have to ensure that the
sine waves are lined up correctly in terms of just how their peaks and
troughs coincide. That phase information isn't shown in the spectrum
above (nor in any spectrum we've considered so far).

Indeed, very often physicists (and scientists in general) don't really
care too much about the phases of the waves that make up a signal;

---

[15] At the moment I am plotting the data in a way so that we only see the most
intense peaks: the frequency spectrum is plotted on a linear scale. There are
other, weaker frequency components also present that we could see if we plotted
the spectrum on a logarithmic scale. Let's not worry about those weaker peaks or
logarithmic scaling for now: the more intense peaks are all we need in order to
see Fourier analysis in action.

often, knowing just the frequencies and amplitudes is enough to get a good handle on what's going on. Phase is a fascinating aspect of the physics and maths, however, and is especially relevant in the context of visual, as opposed to audio, signals (for an image, how the waves line up is key: offset the peaks and troughs and a photo, logo, album cover, etc. can be hopelessly scrambled. We'll see much more on this later in the book).

We'll get back in(to) phase in Chapter 10. For now, let's add some full-on Fear Factory distortion to the "Scumgrief" harmonics and take a look at the waveform and corresponding frequency spectrum . . .

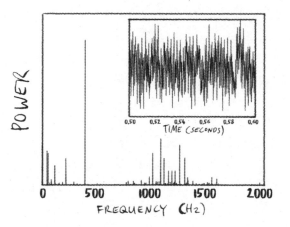

The distorted guitar signal is much more harmonically rich than its clean counterpart.

🔊 You can hear as much by listening to the original (here: https://youtu.be/Hj4ua0ybqRg) and then its distorted counterpart (here: https://youtu.be/f-fRDh_qY_k).

Where do all the extra harmonics come from? They arise simply because of the additional complexity of the distorted signal: we need a greater number of colors from the audio palette to represent the waveform. For one thing, take a look at the portion of the waveform for the "Scumgrief" riff shown in the inset above and you'll see that the distorted version is much more "jagged" than the clean guitar version shown previously—the signal changes on a much shorter timescale as compared to the clean guitar harmonics. You'll recall that there's a reciprocal relationship between time and frequency, so a signal changing on a short timescale is necessarily associated with high-frequency Fourier components.

By filtering the signal in the correct way, those high-frequency components can be reduced (or amplified). Or we could choose instead to amplify the low-frequency components: to boost the bass. Remember, every time you adjust the EQ (or treble and bass) on any piece of audio equipment—from an iPhone to that ginormous amplifier stack we considered in Chapter 2—you're changing the Fourier spectrum. In this case, if we filter out the higher-frequency components (above 500 Hz) of that distorted "Scumgrief" riff, here's what we get. (The filtered version is on the left, and an original, unfiltered, everything-turned-up-to-eleven sample is on the right for comparison.)

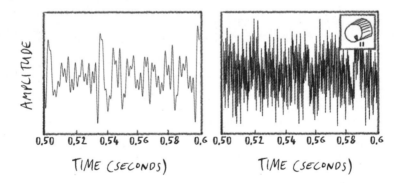

The filtered waveform is much smoother, as we'd expect. A "spiky" signal varies quickly in time and we therefore need high-frequency components in the mix to represent these rapid changes. When we remove those high-frequency components, we smooth out the signal.

Listen to the filtered version here: https://youtu.be/ZLyO5HoCrf4.

Cry Baby

By far the most metal way to do this type of Fourier filtering is to use a wah-wah pedal. Invented in the 1960s, the wah-wah is arguably the most expressive effect available in the battery of pedals and signal processors used by guitarists. A wah filters the guitar signal so that frequencies outside a certain range (or "band") are attenuated. The center frequency of that "band-pass" is adjusted by rocking the pedal back and forth.

When you hear the effect in action, you realize instantly why it's called a wah-wah, and why the most popular version of the pedal is called the Cry Baby. It's a remarkable effect that mimics, to some extent, the sound of the human voice. As such, it's been exploited throughout rock and metal since its invention, with Jimi Hendrix being a particularly enthusiastic early adopter of wah technology. The wah-wah pedal has since been responsible for many landmark moments in metal: the first song on Iron Maiden's eponymous debut kicks off with a wah-laden riff; Black Sabbath's "Turn Up the Night" off *Mob Rules*[16] is driven by Tony Iommi's thrilling wah-effected trilling;[17] Cliff Burton's bass motif in Metallica's "For Whom the Bell Tolls" is a rare example of wah being masterfully applied to bass guitar (although Geezer Butler had of course opened up "N.I.B." with wah-modulated bass a decade earlier); and, when it comes to demonstrating the "vocal" qualities of the wah effect, Steve Vai did it best in the call-and-response opening of David Lee Roth's "Yankee Rose."

---

[16] Although it's generally accepted that of the albums Ronnie James Dio recorded with Sabbath, *Heaven and Hell* is the classic work—and, indeed, is a classic metal album in its own right—I prefer *Mob Rules*. It's a darker, heavier album underpinned by sublime production by Martin Birch.

[17] A *trill* is a rapid alternation between two notes. On guitar it's generally played by following a hammer-on with the fretting hand by a pull-off—the notes aren't picked so the picking hand can take a breather. AC/DC's Angus Young can often be seen with his right hand high in the air while he trills away with his left hand.

Let's bring this chapter full circle and close with a return to the music of *Master of Puppets*. (One might even say we're going back to the front.) Kirk Hammett is a particularly enthusiastic exponent of the wah pedal and uses it to great effect in his solo on "Battery," the opening track of *MOP*. Let's look at the first notes of that solo. Hammett plays what's called a double stop (that's two notes played together—a *diad*) and rocks the wah pedal back and forth to modulate the tone of those notes. I'll do just what Hammett does—but without the wah pedal. Here's the frequency spectrum of those opening notes (which are an A and a C♯ played on the second and third strings but bent in pitch by pushing the strings toward the top of the guitar neck):

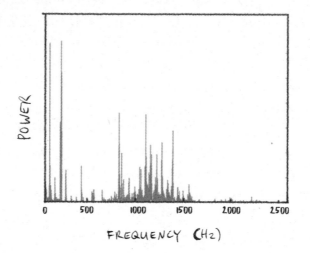

Don't worry about the complexity of the spectrum—there are lots of peaks there, but that's to be expected given that we're playing two notes together and there's a wash of distortion.

But now let's compare the frequency spectra when the same notes are played (in the Hammett style) with the wah pedal switched on, and either depressed so it's closest to the floor or in the "up" position. (I've included the original "no wah" spectrum in the figure to make the comparison easier.)

The band of frequencies that the wah pedal accentuates shifts as a function of the position of the pedal. Or, in slightly more mathe-

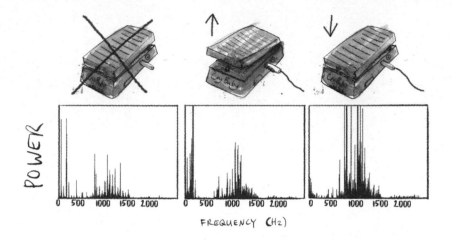

POWER

FREQUENCY (Hz)

matical/physics-y language, the wah pedal modifies the mix of Fourier components in the guitar signal. It's that modification of the Fourier spectrum that is responsible for the wonderfully rich tonal variation provided by the wah pedal.

> **Here's the audio for the opening notes *sans* wah: https://youtu.be/5aHGgql47Co . . .**

🔊 **. . . and as Hammett would prefer: https://youtu.be/ h3Qglotw7Rg.**

So far we've considered metal in the context of the *classical* world. Although we've seen that the wavelengths and frequencies of notes on a guitar string are quantized—they can only have certain values, which is the very essence of quantum physics—a guitar string is *massive* compared to the submicroscopic world in which quantum physics is writ large. This world of the ultrasmall hasn't yet factored into our discussion. That's about to change in a big—yet also very, very small—way.

Chapter 5

A QUANTUM LEAP: THINKING INSIDE THE BOX

I'll be more enthusiastic about
encouraging thinking outside the box when
there's evidence of any thinking going on inside it.

—Terry Pratchett[1]

---

[1] Terry Pratchett, on alt.fan.pratchett.

Heavy metal saved my life.

Well, maybe not my life—but certainly my career. Let me explain.

As an undergraduate physicist at Dublin City University (DCU) in the late 1980s, I found myself increasingly disillusioned. This had absolutely nothing to do with DCU or the program itself: it was simply that I lost motivation. Part of this was certainly due to my focus on music; I was regularly returning home on weekends to rehearse and play in a band. But another reason was that, although I loved the subject, I wasn't sure I loved the work, and increasingly felt that I just wasn't cut out to be a physicist. This all culminated in failing my end-of-third-year exams. Badly. So badly that I had to repeat that third year. And I failed those exams largely because, instead of studying and revising, I dedicated my time to learning songs, writing riffs, working on arrange-ments, and replaying cassettes over and over again in an attempt to deduce just how Gary Moore or Jake E. Lee or Alex Lifeson had played a guitar solo. (Despite wearing out those cassettes with the constant rewind-replay cycle, I *still* didn't get it right.) This was clearly not a clever strategy in terms of my aspirations to a career as a professional physicist.

Or so it seemed at the time. In hindsight, however, failing my exams in third year was the very best thing that could have happened to me. If I'd passed those exams, the very strong likelihood is that I'd have drifted through my final year and graduated with a third-class honors degree or, at best, a lower second—not enough to start a PhD. Instead, failing my exams hit me—in the words of a Judas Priest classic—*"like a batter-ing ram."* It scared the *bejaysus* out of me and made me realize that, yes, I did want to devote my days to the exhilarating if sometimes thankless pursuits of the practicing physicist, and it was time to knuckle down. I graduated with an upper second–class honors degree—enough to begin a PhD program. And during that program, I'd often get irritated at having to leave an experiment to go rehearse with the band: I'd come full circle. I didn't realize it at the time, but there were strong links between metal and the physics I was doing. I just hadn't spotted them yet.

At the heart of my PhD project was an instrument called a scanning probe microscope. An SPM is a microscope like no other: no lenses, no mirrors, no optics. Instead, an image is generated by scanning a sharp probe back and forth across a surface, measuring how the interaction between the probe and the surface changes at each point. But the truly remarkable aspect of the SPM is that if the probe is atomically sharp—which is not as difficult to achieve as it sounds—then we can see not only individual atoms and molecules, we can move them one by one to form structures that have never before been seen in nature. The cartoon below illustrates just how we do this.

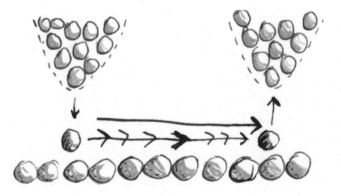

The first example of this type of atomic manipulation was published in the year I started my PhD (and played no small part in my decision to take up a doctoral course) and is seen on the next page. It shows the letters "IBM" spelled out in xenon (Xe) atoms on a nickel surface. Each atom was dragged across the surface "by hand"—well, by a scientist controlling a computer that directed the probe—in a process that took a little less than a day (twenty-three hours to be precise).

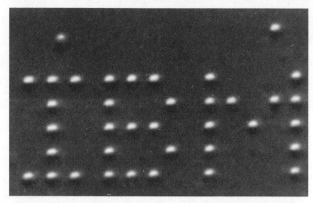

Reprinted by permission from Eigler, D. M., and Schweizer, E. K., "Positioning single atoms with a scanning tunneling microscope," *Nature* 344, no. 6266 (1990). © 1990 by Springer Nature.

Atom by Atom[2]

You might ask how it's possible to not only create an atomically sharp probe, but then to move that probe with atomic-scale precision. At first glance, those seem like insanely difficult tasks. Yet, from some perspectives, they actually involve rather mundane "technology" and materials. (That's not to say that these types of single-atom/single-molecule positioning experiments are easy. Before I'm hunted down by a rampaging mob of scanning probe microscopists, I should stress that atomic manipulation with a scanning probe microscope is definitely *not* an easy task.)

We create an atomically sharp tip by first taking a piece of tungsten wire and electrochemically etching it in a process that is no more involved or complicated than a high school experiment. We dip the wire into a solution of potassium hydroxide (or sodium hydroxide),

---

[2] To my delight, Satan—not the Prince of Darkness himself but the band that was a mainstay of the NWOBHM back in the 1980s—released an album with this title not so long ago. Unfortunately, the lyrics of the title track aren't entirely related to the theme of atomic manipulation with scanning probes (but I'll take the metal allusions where I can find them)!

apply a voltage, and wait. The solution erodes the tungsten and eventually forms a sharp tip.

. . . but not *atomically* sharp. The radius of curvature of the tip (see sketch below) is at the tens of nanometers at best. So how do we get to atomic resolution? Fortunately, the interactions that we exploit to image single atoms (see below) weaken so rapidly with distance that one atom sticking out just a little more than those around it can be all we need.

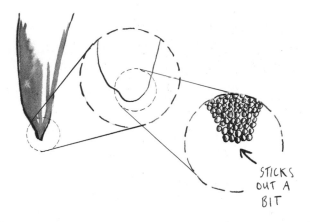

STICKS
OUT A
BIT

Now, sometimes we are lucky enough that this happens when the tungsten tip is placed inside the microscope, but that's rather unusual. More often, we need to apply some tip "treatment."

This is the deep, dark secret of scanning probe microscopists: we will do anything to achieve atomic resolution. First we'll tickle the probe with a small voltage pulse. If that doesn't bring the atoms into "focus," we'll apply a large voltage pulse. And if that doesn't work, we'll gently crash the tip into the surface by a nanometer or three (about ten atomic diameters). Then, starting to become desperate, we'll rather less gently crash the tip into the sample. Getting more irritated by the minute, we'll bury the probe in the surface, drag it across, vibrate it, and then drag it out.[3] All in a bid to nudge a single atom to the fore.

---

[3] This is not to say that it's always as extreme as this. But the scenario I sketch out here is certainly not unusual in the scanning probe community.

These details of probe engineering are all very interesting, I'm sure, but none of this addresses the central question: How is it that the probe microscope can see atoms in the first place? The image is constructed by plotting the height of the probe as it tracks across the surface. The computer controlling the probe is programmed to keep the current between the tip and the sample constant, and the current falls off exponentially as the size of the gap between the tip and the sample increases; this means that the tip moves up and down as it traverses the surface, registering "bumps" in the data. The stream of numbers representing the probe height is converted to an image by, for example, making the lowest height black, the highest position white, and scaling in shades of gray between those two extremes.

In the original—and still very popular—version of the scanning probe microscope, the current that senses the surface is created by a purely quantum physics effect known as tunneling. Once again physicists have borrowed a word from everyday language to describe a quantum phenomenon, but this tunneling is like nothing at all we encounter in our macroscopic, classical, everyday world. (Paradoxically, that's not to say that tunneling isn't an essential process in the world around us . . . you'll see what I mean.)

Consider the scenario shown on the next page. In the macroscopic, classical, everyday world around us, you know what'll happen if the cable connecting the guitar and amp is cut in two: Metallizer's guitarist will be far from impressed that his visionary guitar solo was cut off in its prime and is likely to berate the road crew, informing them loudly that unless his connection is reestablished *tout de suite*, dammit, some heads are gonna roll.[4] Mr. Volumus is having to contend with a catastrophic open-circuit situation. The electrical current that usually carries his flurry of notes to the amplifier is stymied by the break in the cable. Only when the two ends are reconnected will the current flow—and the shredding—resume.

------

[4] Metallizer's lead guitarist, Max Volumus, is so exceptionally old-school that he refuses to use a wireless connection.

That's the situation in Metallizer's (and our) macroscopic world. But it's not what happens in the nanoscopic world of atoms and molecules. The sketch below also shows what we'd see if we could zoom in to the severed ends of Max Volumus's guitar cable and look at the atomic level detail. What's remarkable is that in the quantum world, we don't need the ends of the cable to join up so that a current can flow. In the quantum world, an "open circuit" isn't really an open circuit at all; the gap in the cable is just an obstacle. Electric current—electrons, in other words—can still flow, despite the lack of a conduit for the electric current! It's rather like putting a blockage in a hosepipe, or tying a tight knot in it, and finding that water still flows out the end.

The gap between the ends of the wire can be as large as the width of three or four atoms, and still electrons will flow. Now, you might quite reasonably argue that "as large as the width of three or four atoms" sounds rather oxymoronic—isn't an atom extremely tiny indeed? If we were talking about the macroscopic world, then I'd agree—an atom is a fraction of a nanometer across; there are roughly thirty million of them spanning the width of Volumus's guitar cable. But, in the quantum world, three or four atomic diameters can be a very large separation indeed, especially given the scale of a chemical bond.

Chemical bonds are an essential component of all matter—they're the electron glue that holds everything together. What happens at the most fundamental level when we cut Volumus's cable with a pair of scissors or wire cutters? We break chemical bonds. However, as long as the break is no more than a few atoms across, the bonds in the cable don't need to be repaired; at the quantum level, the electron pipe can have a break in it, and electrons will still flow around the circuit.

As if that weren't enough, the electrons don't lose any energy when they do this. They encounter a barrier to their motion and yet they don't need any additional energy to circumvent that barrier! This is bizarre and runs entirely counter to the physics that works in our everyday life. Consider the scenario sketched below. Metallizer's hapless tour bus driver is less than entirely dependable and has managed to get lost en route to the band's next gig. Badly lost. At 4 AM they find themselves at the foot of a narrow mountain pass. To continue on their way, they need to follow that long and winding road over the mountain—and that is going to cost energy.

But now imagine shrinking Metallizer, their driver, the bus, and the mountain down to the atomic scale—to nanoscopic dimensions. What's remarkable is that this imagined tiny bus doesn't need any additional energy to travel beyond the mountain—it can simply pass straight through and continue on its way, as if by (black) magic. This is quantum tunneling in action. A particle—in this case the nanobus— can pass a barrier (the nanoscale mountain) without requiring any additional energy to do so.

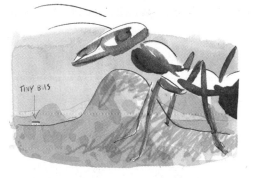

A tired and emotional Metallizer makes it to the next venue (and promptly sacks their bus driver). During their gig, the obligatory circle pit opens up, providing a helpful opportunity for another macroscopic analog to quantum tunneling. Imagine our mosher friend, Marsha, from Chapter 2, trapped at the center of the pit. Unless she expends a great deal of energy, she is not getting out of there anytime soon—she'd have to force her way out through the mass of bodies circulating around her. Shrink the mosher and the surrounding pit down to the nano- scopic scale, however, and it's a different story. Mini Marsha can pass straight through the wall without expending any energy at all.

Macro Marsha   Nano Marsha

— TO THE BAR!

AAAH MY BEER IS EMPTY!

Why doesn't this tunneling effect happen in the visible world? It's all a matter of scale. Only in the world of the ultrasmall are these quantum effects writ large. In the example of the cut cable, the barrier to electron flow is the gap in the cable. We can think of it in terms of the Metallizer tour bus trying to cross a gorge, rather than a mountain. In the macroscopic world of classical physics, the bus needs a tremendous amount of kinetic energy if it's to stand a chance of making it to the other side; shrink everything down to the quantum level, however, and the bus can cross that gorge at a cruising pace without even revving its engines.

The probability for tunneling goes down very quickly indeed as that gap/barrier gets larger. It's an exponential relationship: separate the ends of the cable by an additional atomic diameter (say, 0.4 nm) and the probability that an electron can tunnel across the gap decreases by about a factor of 10,000 (give or take, depending on the size of the atom). Widen the gap by the diameter of another atom and the probability drops by another factor of 10,000. In other words, by moving the ends of the wire apart by a little less than a nanometer in total (0.4 nm + 0.4 nm = 0.8 nm), the probability of an electron making it across the gap goes down by a factor of 10,000 × 10,000, or 100 million.

By the time we've reached the size of a virus—let's take the polio virus, which is about 30 nm across, as an example—the probability of an electron tunneling across a gap that "large," or a barrier that wide, is mind-bogglingly small. Once we reach length scales typical of the world around us—the length of a bus, the height of a mountain, the width of a human body (in or out of a circle pit)—the probability for tunneling is, if you'll excuse my paraphrasing of the late, great Douglas Adams,[5] smaller than the smallest thing ever, and then some.

But it's not zero.

The graph of this probability is an exponentially decaying function.[6] And an exponentially decaying function, as shown in the sketch on the next page, only reaches zero when its "input" is negative infinity. In short, unless the ends of the cable are separated by an infinite distance, there's still a possibility of an electron tunneling. In other words, if we were to take one half of the cable to a planet in a distant star system—let's say Proxima Centauri, which is 4.25 light-years away, i.e., about 25 million, million miles away (or 40 million, million kilometers for those who prefer scientific units)—and leave the other end on Earth, there's still a nonzero probability that the electron might tunnel!

---

[5] Author of my favorite five-book "trilogy," *The Hitchhiker's Guide to the Galaxy.*

[6] A fancy way of saying it decreases exponentially. And that, in turn, is just a fancy way of saying that the decrease is proportional to the size of the quantity of interest (in this case, the current). As compared to a linear decrease, an exponential falloff is much, much faster. For every tenth of a nanometer the tip is moved away from surface, the current drops by almost a factor of ten. If the tip-sample separation increases by just the width of an atom, the current drops by a factor of a thousand. By the diameter of two atoms, and the current decreases by a factor of a million!

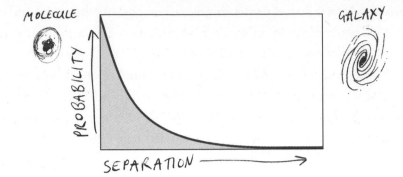

Now, before we get too excited, "nonzero" means nothing more than the probability isn't absolutely zero. That doesn't mean, however, that there's any influence at all, in any way, on anything we can see, hear, touch, taste, or smell. Or on anything we could ever scientifically measure.[7] Nonetheless, there's a great deal of fashionable "woo" out there that suggests precisely the opposite: that the phenomenon that does away with barriers on the nanoscopic scale means we're all part of one integrated, holistic universe and, in a mystical and magical way, are all interconnected and at one with each other. Because quantum.

However much that holistic universe might sound like a rather nice one to live in, quantum physics tells us absolutely nothing of the sort.

Here's a typical assertion from woo-meister extraordinaire Deepak Chopra:[8]

---

[7] I need to stress that this is not simply a question of devices and instruments not being sufficiently sensitive today, and that if we just wait long enough for future advances in technology we'll be able to detect these effects. It's that the effects in question are totally, utterly, and completely washed out by the environment around us. It's like if we whistled here on Earth and then expected to be able to measure an effect on the orbit of that planet in the Proxima Centauri system.

[8] Jordan B. Peterson, professor of psychology at the University of Toronto and author of *12 Rules for Life*, a set of trite aphorisms dressed up as "profound" commentary on societal mores and personal morality, has parroted the nonsense Chopra spouts here regarding quantum possibilities. If you'll excuse my own recycled aphorism, a little knowledge, particularly outside one's own discipline, is indeed a dangerous thing.

In the quantum realm there are no fixed objects, only possibilities . . . In the quantum realm, everything is interwoven and inseparably one . . . Your essential state is a field of infinite possibilities . . . Your body is inseparably one with the whole universe . . . Your physical body is capable of taking a quantum leap from one biological age to another without having to go through all the intervening ages in between.[9]

No.

Really, no.

Your body is not "inseparably one with the whole universe." Nor, of course, is it possible for your body to magically "tunnel" from twelve to sixteen years in age, handily skipping the customary awkward phase. Chopra is mangling quantum physics in the most atrocious manner here—and is wildly extrapolating from the atomic (and subatomic) level to the macroscopic world at large. This disregard for the importance of different length scales—to distinguish the macroscopic world around us from the microscopic and nanoscopic (and subnanoscopic) domains—is a defining feature of quantum woo. And we'll see in Chapter 6 just how dramatically our world of classical physics differs from the quantum regime.

Does weird and wacky stuff happen at the quantum level? Yes, of course. And that's because, despite a century of effort, there's still an awful lot we don't understand about the quantum world. But there's a great deal we *do* understand. And one thing we know full well is that, as Lisa Randall, professor of theoretical physics at Harvard University, pithily puts it: "There are a lot of mysteries about quantum mechanics, but they mostly arise in very detailed measurements in controlled settings."[10] The origin of the vast majority of the quantum woo propagated by Chopra,

---

[9] From *Grow Younger, Live Longer*, published in 2001.

[10] See Robert Irion's interview of Lisa Randall in *Smithsonian Magazine*, December 2011: www.smithsonianmag.com/science-nature/opening-strange-portals-in-physics-92901090/.

and others like him,[11] is a deep confusion about the relationship between what happens at the nanoscopic level and the physics of the everyday world around us.

This is not to say that the principles of physics and maths that we see in action "up here" are completely *un*related to what's happening "down there." Far from it. Indeed, the entire premise of this book is to highlight those connections between the everyday—well, *okay*, "everyday" in a very metal sense—and the rather less mundane quantum world. Just because we can draw those parallels doesn't mean that we can assume the physics of the ultrasmall scales up. As we are beginning to glean in some detail, the physics and maths behind all the riffs and solos we know and love so well can indeed be used to describe and understand what's happening at the quantum level. But does that mean that a guitar string is just a scaled-up version of an electron or a proton? No, of course not. Many of the same mathematical principles rule the behavior of the guitar string and the electron, but it would be a blunder of galactic proportions to assume that one is just a giant version of the other.

LAB IN A LORRY

The following story might be helpful in explaining what I mean about drawing a distinction between the parallels that exist in the physics/maths of very different systems and the physics of the systems themselves. From some perspectives this is arguably a subtle difference. From others, as we're about to see, it's anything but.

A number of years ago I helped out with a great project set up by the UK Institute of Physics called "Lab in a Lorry." This involved a touring lab—in a lorry,[12] natch—that visited schools all across the country (and, indeed, even in other countries). There were four experiments on board—or there

---

[11] "The world is actually made of potential, and that potential is actualized by consciousness . . ." Prof. Peterson or Dr. Chopra?

[12] For those across the pond: a truck.

were when I helped out, in Nottingham—and after a short explanation from the Lab in a Lorry team, the students had time to perform them on their own. The experiments were not especially complicated, but established extremely important physics principles. And it was a lot of fun.

One experiment on the lorry involved an exploration of why the sky is blue. Fundamentally, this is due to the scattering of light of different wavelengths by molecules in the atmosphere. Different wavelengths mean different frequencies, thus different energies and different colors: more of the higher-energy (shorter wavelength) blue light streaming in from the sun (among all the other colors making up the spectrum) gets scattered about by the molecules of our atmosphere; hence the sky appears blue. When the sun moves into our line of sight as it approaches the horizon at sunset, the light must pass through more of the atmosphere to reach us, scattering more and more of the blue light away until what we see is the red left over.

To demonstrate this effect, which is called Mie scattering, we used a long glass cylinder filled with water, lit from below with a flashlight. By adding a very small amount of dishwashing liquid (not enough to color the water, and from a company whose product, it was claimed, also had particularly beneficial effects with regard to hand softening), we create something that simulates our atmosphere full of scattering molecules. Our lit-up tube of soapy water looks sky blue, and when viewed straight on from above, the light from the flashlight appears red—like a mini sunset.

I enjoyed chatting about this experiment with the students who came on board the lab. Often with primary schoolchildren, however, the analogy could be lost (despite my best efforts), and I'd then have to spend quite a bit of time explaining that the reason the sky was blue was *not* because it had dishwashing soap in it. And no, we didn't need to worry about how the detergent got into the sky, because that's not what was causing the color. And no, the reason we have rainbows is not because the washing-up liquid in the sky makes bubbles when it rains. And, yes, bubbles *do* have rainbows too, but . . .

The quantum woo that is so prevalent "out there" suffers from a very similar fundamental flaw in reasoning: while there may well be parallels in the physics of the quantum and the classical, we can't assume that this means there's a one-to-one correspondence between everything that happens in the world of leptons and that of lorries.

Heavy Metal Cowboys

Let's get back to Metallizer's tour bus dilemma, chasm version. You'll recall that, in the real world, the only way that their battered and bedeviled mode of transport can make it across the gorge is to pick up enough energy to cross the divide. Shrink everything down *Fantastic Voyage* or *Innerspace* style,[13] however, and the bus suddenly makes it across with no additional energy required.

How is that possible?

You might suspect, on the basis of what we've covered in the opening chapters, that the reason the mini Metallizer bus can make it across the gorge is going to have something to do with waves. And indeed it does. But we're going to see that these are waves like no others. They're a particular type of wave that physicists still don't quite *fully* understand. We can do exceptionally accurate calculations based on this wave model, but we're still struggling with precisely how to interpret what those waves *mean* in certain circumstances. And this isn't just some esoteric problem with no relevance outside the physics lab: understanding those waves is at the very core of understanding how our universe is constructed.

The reason that the nanoscopic Metallizer bus can make it across the gorge is that down at the quantum level—in the world of atoms, molecules, electrons, photons, and subatomic particles—all matter has wavelike characteristics. And like any wave, this particular wave is not

---

[13] Miniaturization at the movies. *Doctor Who* also did a *Fantastic Voyage* "homage" where the Doctor and companion(s) were shrunk down to fit inside a Dalek.

fixed in a given position; it spreads out, just like the water ripples on a pond. Like this:

Image courtesy Professor Michael Crommie and Don Eigler, IBM Almaden Research Center.

That picture is my favorite image in all of science. More than twenty years after I first saw it, it still grips me.[14] It's a ring of forty-eight iron[15] atoms—literally a heavy metal circle—painstakingly arranged one by one using a scanning tunneling microscope (a kind of scanning probe microscope) to form the structure you see. Each atom has been dragged across the surface and put in its place as we described before, building the circle one atomic block at a time. This all happened at a temperature

---

[14] This image is in black and white here, but it is often shown as brilliantly colorful. However, an SPM image is not like one captured by a camera, a telescope, or a conventional microscope—there is no "true" color. Remember, the image is constructed by plotting the height of the probe as it tracks across the surface. The stream of numbers representing the probe height is then converted to an image not unlike a topological map, by, for example, making the lowest height black, the highest position white, and scaling in shades of gray. Or shades of red, blue, or whatever you choose. In other words, an atom can be whatever color we like in an SPM image.

[15] Iron, along with copper, tin, gold, and silver, is classified as a heavy metal due to its relatively high atomic weight.

of roughly four degrees above absolute zero—a very, very chilly minus 269 degrees centigrade—and in an experimental chamber holding a vacuum roughly comparable to that found on the surface of the moon. The exceptionally low temperature is required because otherwise the iron atoms will pick up thermal energy, i.e., heat, from their surroundings (including, in particular, the copper surface on which they're resting) and start to jiggle around. Even raising the temperature by twenty degrees or so (to a wonderfully balmy –250°C) is enough to cause the circle to fall apart[16] as diffusion of the atoms sets in.

The ultrahigh vacuum, on the other hand, is required so that the sample surface doesn't get contaminated, corroded, or corrupted by atmospheric gases. It's not always the case that such vacuums are required—some surfaces are remarkably inert even under atmospheric conditions—but the vast majority of samples will have their surface atoms "scrambled" by exposure to atmosphere.

It took many hours of careful experimentation to build that atomic circle, or, to use the description coined by the scientists who created it, the "quantum corral." It's an astounding feat of atomic-scale engineering (carried out by Mike Crommie and colleagues at the IBM Almaden research center[17] and published all the way back in 1993): an entirely artificial arrangement of atoms, never before seen in nature or science (or, for that matter, *Nature* or *Science*), assembled under computer control.

But that's not the good bit. And that's not the reason I love that image. The reason I love that iconic quantum corral picture is because of what we can see *inside* the circle. I described that pattern of concentric rings in terms of water ripples on a pond. A much better analogy, however, is with the pattern that forms in a cup of coffee.

---

[16] *"Break the circle and stop the movement,"* sang a certain Ronnie James Dio on a classic Sabbath track. This is precisely the opposite. The movement of the atoms breaks the atomic circle.

[17] Professor Michael Crommie now heads up a world-leading scanning probe microscopy/nanoscience research group at the University of California, Berkeley.

Wake Up and Smell the Quantum

You're sitting on a train in a station and it's gently resonating, about to depart. (Actually, if you're in the UK, it's likely to be less a gentle resonance and more akin to a good old bone-rattling—nervous—shakedown.) The entire carriage vibrates, and when you look down at the coffee in front of you, an intriguing pattern of concentric circles has formed.

Compare and contrast this mental image with the image on page 97. What's remarkable is that the pattern you see in your coffee cup[18] is identical to that seen in the quantum corral. Not just similar. *Identical.* And that's despite the fact that the diameter of the coffee cup is about six million times larger than the atomic corral. *And* that there's not a hint of coffee to be found in the ultralow temperature, ultrahigh vacuum environment in which the atomic circle was constructed. *And* that coffee is clearly a liquid, while the iron atoms and the copper surface on which they're supported are solid.

So if everything about the two systems—coffee cup vs. quantum corral—is so different, why is the pattern the same?

Well, what do these two very different systems have in common?

Yep, the circle. The only thing that connects the quantum corral with the coffee cup is the circular symmetry. *This* is why the pattern

---

[18] Coffee isn't perhaps the first beverage one might associate with metal but Anthrax, for one, certainly expressed their appreciation for a good cup of joe in their song titled "Cupajoe" from *Volume 8: The Threat Is Real* (Victor Talking Machine Company, 1998, the lyrics of which include the word "cupajoe" repeated ten times in a row, followed by *"Black and strong! Black and strong!"* Anthrax has not always been concerned with exceptionally accomplished lyricism, it must be said). And there's another coffee reference I can't let pass. Henry Rollins might not be best pleased to be cited in a book on metal—in his 1994 memoir, *Get in the Van*, he describes metal as follows: "I would love to play with fucking 'heavy metal' bands more often. It was fun crushing them. It's all lights and makeup. What bullshit."—but I'm a fan of Rollins and I have to flag up Black Flag's classic "Black Coffee" from *Slip It In* (SST Records, 1984): *"Drinkin' black coffee, black coffee, black coffee, starin' at the walls."* (Rollins's criticisms of metal notwithstanding, he *is* a huge Black Sabbath fan . . .)

is the same. Completely different physics, but identical mathematics! This is the type of mathematical elegance that gets physicists all het up: a connection that spans the ultrasmall to the macroscopic, and beyond. Patterns that appear across vastly different length scales. Discovering these types of connections is as exciting to a physicist as hearing the riff to "Smoke on the Water," or "Highway to Hell," or "Raining Blood," or "The Number of the Beast" for the first time.

The pattern in the coffee cup is a standing wave, just like the standing wave on a guitar string, except for one key difference: the boundaries confining the wave. The guitar string is, in essence, one-dimensional. We can treat it like a line. Of course, in reality that line has a certain thickness, but mathematically, we can consider the string a one-dimensional object.[19]

The surface of the coffee, however, is two-dimensional: it's not a line, it's a flat surface—it has length *and* breadth. And just as traveling waves running up and down the length of the guitar string can constructively and destructively interfere to form a standing wave, so too can the waves on the surface of the coffee (which are excited by the cup being vibrated). Those waves are confined, like those on the guitar string, and interfere to form a standing wave. But this time the pattern is considerably different. Instead of sine waves, the result is something called a Bessel function.[20] For our purpose, we don't need to consider Bessel functions in any detail. What we do need to realize is that the same pattern of ripples you see on your cup of joe pops up everywhere in nature.

That a standing wave can form on the surface of the coffee is perhaps not too surprising. It's a liquid; liquid can make waves (as anyone who's had even the briefest of encounters with a beach will have ascertained);

---

[19] More on dimensions in Chapter 8, I promise.

[20] . . . but remember that *any* pattern can be built up from the right choice of sine waves. And a Bessel function is no different.

and those waves can reflect. All the ingredients are in place for the for-mation of a standing wave.

But just what is it that's *waving* in the quantum corral? The iron atoms making up the circle are fixed in place and are analogous to the "walls" of the coffee cup. The copper atoms making up the surface on which the Fe atoms have been so carefully rearranged are also immobile. Where does the wave come from?

The standing wave seen in glorious not-so-Technicolor on page 97 is a wave of electrons at the surface. We're seeing how the distribution of electrons is affected by being trapped within the circular corral of sur-rounding iron atoms. We're seeing, right before our eyes, the wavelike nature of matter. Electrons are behaving like waves, seemingly spread-ing out in space, bouncing back and forth, and faithfully reproducing the coffee cup pattern at the quantum level.

Physicists refer to the corral more generally as a *potential energy well*: the electrons are confined and can't escape, much like Marsha is trapped at the bottom of the well on the next page. In Marsha's case there's a difference in her potential energy at the bottom of the well and that she'd have if she were standing on the ground beside it due to gravity (remember that the formula for her potential energy is *mgh*, with the *h* standing for height). The electrons in the corral, however, are much too tiny to care about gravity. Instead they're confined by another type of energy difference that is due not to gravity but a force governing interactions between electrons and atoms.[21] They're similarly trapped in a well: a quantum well.

---

[21] Ultimately, the source of this energy is the electromagnetic force, one of the four fundamental forces in nature alongside gravity, the strong nuclear force, and the weak nuclear force. I much prefer columnist Dave Barry's sixfold classification of the fundamental forces in nature, however: "Magnetism is one of the six fundamental forces of the Universe, with the other five being Gravity, Duct Tape, Whining, Remote Control, and The Force That Pulls Dogs Toward the Groins of Strangers." (Barry's classic definition first appeared in the *Miami Herald*, June 1997.)

Those of you familiar with the happenings at a certain massive underground laboratory straddling the Swiss-French border, where physicists and engineers spend their time crashing the fundamental building blocks of matter into each other,[22] might at this point raise a very pertinent question. Aren't electrons particles? Indeed, isn't there an entire branch of physics known as particle physics? Why don't the electrons in that quantum corral image appear as little "billiard ball" particles? Why are they wavy?

And therein lies the conundrum at the core of quantum physics known as *wave-particle duality*. The building blocks of matter have a dual character. Sometimes they behave for all the world as if they're solid little billiard balls. Yet look at them in a slightly different way and they appear to be diffuse, cloudy waves: slippery and difficult to pin down. The latter is exactly what we see inside the quantum corral. The electrons at the copper surface form what physicists call a *delocalized* wave pattern.

---

[22] Otherwise known as the European Organization for Nuclear Research (CERN), which houses the world's largest particle accelerator. And yet again, we find a metal-physics connection here. In 2015, a CERN physicist took data from the Higgs boson discovery at CERN and turned it into a metal song: www.youtube.com/watch?v=SXEnDM3hydM.

They're spread out across the surface: we can't assign them specific positions as we could billiard ball–like electrons. We can't *localize* them.

Now that would be weird enough if it were the case that matter itself spreads out like a wave. But that's not what happens.[23] It's not as if an electron gets put on a nanorack and is stretched out.

Instead, the wave we see inside the corral is a map of the *probability* of finding an electron. This might seem like a bizarre concept, but handily we can draw a parallel with something more tangible. For example, here's a map of the UK and Ireland showing how Metallizer's ticket sales worked out for their most recent tour.

---

[23] No, that of course would be too simple by halves. Thanks, Nature . . .

The different shades of gray on the map represent variation in the number of tickets sold: the darker the shade of gray, the greater the ticket sales in that particular region. Compare this to the image of the quantum corral. The shading there represents how the number of electrons flowing between the tip and the sample changes from place to place on the surface. It's a map of the probability of finding an electron.

In other words, Metallizer's ticket sales graphic is, in essence, a map of popularity. And the quantum corral is . . . a map of probability.

So it's not that the electron itself *is* a wave; it's that it behaves *like* a wave. In principle (if we had enough sensitivity in the measuring equipment), we could register each individual electron[24] as it flowed from the corral to the probe. As it stands, we measure an electrical current—a flow of electrons between the tip and the sample—and the number of electrons we measure depends entirely on that probability map. It's like this:

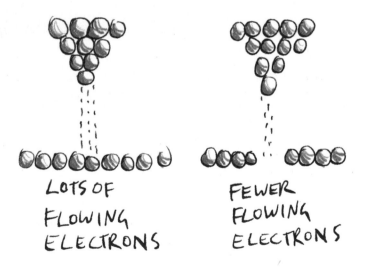

LOTS OF FLOWING ELECTRONS

FEWER FLOWING ELECTRONS

[24] In other experiments, this has been done—it's called single-electron charging. In later chapters we'll see how the wave pattern for a beam of molecules has been measured one molecule at a time. Similarly, it's experimentally possible to build up the probability wave for electrons on a particle-by-particle basis.

In other words, the wave defines where we'll find electrons. But the electrons themselves aren't waves. Yes, I know that's confusing. Deeply confusing. And don't think for one second that physicists are entirely comfortable with this wave-particle duality thing. We just get used to it. There are still many questions to be answered about what's actually happening at the quantum level: What does that probability wave *really* represent? What is it telling us about how the universe works at the most fundamental level? Is it somehow a question of what we know rather than what's "out there"?[25]

But if we put these thorny philosophical issues to one side, the remarkable thing is that the wave theory *works*. And it works exceptionally well.

Music of the (Nano)spheres

The circle of atoms that forms the quantum corral is of course just one way that atoms on a surface can be arranged. A rather more straightforward pattern is to simply line up the atoms in a row. This is exactly what Wilson Ho and his colleagues at the University of California, Irvine, did back in 2005, forming another type of "corral" for the electrons (and their associated waves). They assembled chains of gold atoms of different lengths, ranging from a gold dimer (a pair of bonded atoms) all the way up to a chain twenty atoms long. A selection of those chains is shown on the next page where we have, in order of appearance from left to right, Au_1 (a single gold atom), Au_2 (a gold dimer), Au_3 (three atoms—a linear trimer), Au_4, Au_5, Au_8, Au_{11}, Au_{14}, Au_{17}, and Au_{20}.

---

[25] For a gripping appraisal of the deep issues surrounding interpretation that continue to plague quantum mechanics, read Philip Ball's "Beyond Weird" (The Bodley Head, London (2018)).

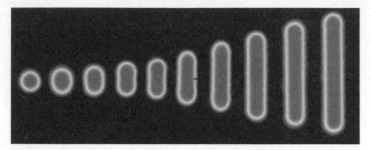

Image reprinted with permission from Wallis, T. M., Nilius, N., and Ho, W. *Physical Review Letters* 89, no. 23 (2002). © 2002 by the American Physical Society.

In other words, Professor Ho and his team created nanoscopic "strings" on which electron waves could be confined. The gold chains you see on the previous page are the atomic equivalents of guitar strings. Note that they're also one-dimensional—a single atom wide. (Well, apart from the lone Au atom on the farthest left. That's *zero* dimensional.[26])

What's amazing is that the probability wave for an electron along each of those nanostrings is identical to the standing waves on a guitar string we covered in Chapter 4. It's the same concept, just shrunk down to the atomic limit. The electron behaves like a wave. Waves on the atomic string are confined. That means those waves are reflected at each end of the chain and form a standing wave. Waves on a guitar string follow the same pattern as the probability of finding an electron in an atomic chain.[27] (And, as we saw for the coffee-corral crossover, the only thing common to both systems is the symmetry—in this case, linear.)

---

[26] More on just what this means later.

[27] I should be careful here or my fellow physicists will run me out of town. Technically, I should talk about probability *density* rather than just probability. Many scientists often get too hung up on technicalities—you might even say they indulge in a little technical ecstasy—but there's a time and a place for minutiae of that type.

It's clear that when we're at the quantum level, the very last thing we want to do is think *outside* the box. Confinement, entrapment, and incarceration are all part and parcel of the quantum world: by trapping electrons in ways of our choosing we can radically change how they behave.

But if atoms and electrons—and all their sibling quantum particles—are associated with waves of probability, and we ourselves are made of those particles, why is it that this "wibbly-wobbly" nature of the universe isn't visible around us? Where do the waves go?

Chapter 6

GIANTS OF THE INFINITESIMAL[1]

When we will look around and see
We will affect the energy . . .
Omne est vigor

—from Mark Jensen's lyrics for Epica's
"The Quantum Enigma"[2]

---

[1] Unlike the vast majority of the other chapter titles in this book, "Giants of the Infinitesimal" is not taken from a metal lyric, nor a song title, nor an album title (although it'd work damn well as any of those). It's instead the name of an arts-science collaboration in which I participated a number of years ago, driven by the fantastically talented duo of Tom Grimsey and his colleague Theo Kaccoufa. The wonderfully evocative "Giants of the Infinitesimal" title was Tom's, and it later became the title of an inspiring book he worked on with the science writer Peter Forbes. Tom very sadly passed away, at age fifty-four, in 2014 after suffering from cancer. I'm using the "Giants . . ." title here both because it's exceptionally apt for the themes of this chapter and also in tribute to Tom. He was an inspiration to artists and scientists alike.

[2] *The Quantum Enigma*. Nuclear Blast, 2014.

"Your Entire Life Is an ILLUSION: New Test Backs Up Theory That World Doesn't Exist Until We Look at It"

"Near-Death Experiences Occur When the Soul Leaves the Nervous System and Enters the Universe, Claim Two Quantum Physics Experts"

"Quantum Physics Proves That There IS an Afterlife, Claims Scientist"

"20 Examples of the Mandela Effect That'll Make You Believe You're in a Parallel Universe"

ARRRGGHHHH! Make. It. Stop.[3]

Those first three headlines are taken from the *Daily Mail*, a tabloid also known in the UK as the "Daily Fail" due to what's perceived by many as a worrisome lack of what might best be called journalistic rigor.[4] Allegedly.

---

[3] I'm not usually a fan (nor a proponent) of This. Style. Of. Writing, but I'm making a rare exception because the cognitive pain I'm experiencing fully justifies the staccato response.

[4] Having said that, even the *Daily Mail* has been critical of Deepak Chopra's pseudoscientific quantum woo nonsense. In an article headlined "Not So Profound Now!" the *Mail* reported favorably on a study by Gordon Pennycook and colleagues at the University of Waterloo entitled "On the Reception and Detection of Pseudo-profound Bullshit" (*Judgment and Decision Making* 10, no. 6 (2015): 549–63). Here's a representative conclusion of their work, taken directly from their paper: "Profundity ratings for statements containing a random collection of buzzwords were very strongly correlated with a selective collection of actual 'Tweets' from Deepak Chopra's 'Twitter' feed (r = .88–89)." I'm not making a judgment on the accuracy of their findings here, but I must admit that I found the call for a "reliable index of bullshit receptivity" elsewhere in their paper to be rather compelling. No more Chopra from hereon in, however. Promise.

You can probably make a good guess as to the source of the "20 Examples of the Mandela Effect . . ." headline. It's a social media and entertainment company that produces many, many lists (and lists of lists). I hadn't encountered the "Mandela effect" previously, so I've got BuzzFeed—for it is they—to thank for bringing that particular nugget of quantum woo to my attention. The Mandela effect takes its name from the entirely mundane observation that there are apparently quite a few people out there who think that Nelson Mandela passed away rather earlier than he did (in the 1980s instead of 2013). But instead of taking this lapse of memory at face value, some have decided that the so-called Mandela effect provides a compelling piece of evidence for the parallel worlds theory of quantum mechanics: that this collective ignorance is good reason to suspect that we're now living in a different reality. (You don't know how much of an effort it took to write the preceding sentence while also gently banging my forehead against the keyboard.)

BuzzFeed "helpfully" provided a list of nineteen other examples of where our universe has apparently split off into parallel realities, including such stunning revelations as: Curious George never had a tail, Kit Kat (the candy bar) doesn't have a dash in its name, and C-3PO isn't made entirely of gold.[5]

There's a myriad of misinformed memes about quantum physics circulating out there that rival that BuzzFeed list for inanity. It's understandable to some extent, because quantum physics is conceptually difficult and there's an awful lot we still don't understand. We scientists also have to shoulder a portion of the blame, because we're not always as careful as we should be to explain to nonscientists that what's occurring "down there" doesn't directly scale to what's happening "up

---

[5] Actually, that last one *is* quite strange. I had to go back and check and, feck me, they're right. Clearly an alternative reality has indeed appeared!

here."[6] Although the molecules, atoms, electrons, and other subatomic particles of which we, and everything around us, are made behave according to quantum mechanics, we nonetheless live in a world of classical physics. We humans aren't waves. We don't refract. We don't diffract.[7] And no matter how chaotic the mosh pit might get, we certainly don't scatter off each other in the same way waves do, rippling and decaying into the distance.

BECAUSE! QUANTUM! PHYSICS!

The next time you're down at the pub, or at a party or a gig (or all three simultaneously) and someone attempts to tell you that we're all part of one interconnected consciousness, that we can influence molecules with our minds, that water has memory, that crystals generate positive energy, and/or any other similarly bonkers claim *because quantum physics*, ask them the following simple question: When was

---

[6] On this point, I have some qualms about how the research area increasingly known as quantum biology is frequently portrayed in the popular science media. Yes, there are some very interesting quantum effects happening in biological systems, including some rather unexpected recent discoveries related to quantum coherence (i.e., the extent to which quantum waves remain in phase and aren't scattered by their environment). The problem, however, is that these effects are still happening on exceptionally short-length scales. This isn't to say that the bioscience in question isn't exciting, but in our excitement to convey the weirdness of the quantum, we scientists sometimes neglect to highlight that it's Newton's theories that still hold sway in the big, classical world around us.

[7] There's lots more coming up on diffraction and the interference of waves that gives rise to the effect. For now, suffice it to say that when a wave encounters an obstacle that is roughly the same size as its wavelength, it spreads out—it *diffracts*—in a fascinating manner that is intimately connected with many aspects of quantum mechanics. Diffraction is also an effect you'll have encountered in a heavy metal context. That rainbow of colors you see on the business side of your Priest, Maiden, Slipknot, Testament, or Leppard CDs? That's diffraction of light in action. (If streaming music continues to grow at the rate it did while I was writing this book, I'll soon face an audience of blank faces in my undergraduate lectures when I mention a CD or a DVD as an example of diffraction.)

the last time you diffracted when you walked through a doorway?

Don't be surprised if they brush you off by mumbling something about not thinking on the "correct level" and that you're "trivializing" or "oversimplifying" what they believe. But you should push the point. If their argument is that quantum mechanics scales directly from the world of the ultrasmall to the macroscopic world around us, then it must follow that its most fundamental aspect—that matter is imbued with wavelike characteristics—should similarly be at play on human-length scales. If the weird quantum mechanics of atoms, molecules, and subatomic particles indeed operates on everyday scales, then we should see evidence for the wave properties not just in the experiments we do but in every human activity. But it doesn't. And we don't. We're not walking wave functions.[8]

During what might be described as the adolescence of quantum mechanics in the 1920s, the historian-turned-physicist Louis de Broglie wrote down a breathtakingly simple equation, requiring no mathematics beyond basic algebra, that tells us how to work out the wavelength of a matter wave—it can be used to debunk the "walking wave function"[9] nonsense in just a line of high school maths. Broglie's equation goes like this:

$$\lambda = h \, / \, p$$

---

[8] This suggestion has been made, not by a practitioner of woo, but by a serious academic, Professor Alexander Wendt, Ralph D. Mershon Professor of International Security at Ohio State University, in a book published in 2015 by Cambridge University Press: *Quantum Mind and Social Science*. While Wendt does a good job of explaining key advances and conceptually tricky concepts in quantum physics, I'm afraid that I have to take issue with his core concept. According to his Wikipedia page, not only is Wendt a prestigious political scientist, but he's a massive Metallica fan. As a fellow metal fan, I'll send Prof. Wendt a copy of this book when it's published.

[9] Diffract on over to the "Because! Quantum! Physics!" box to understand the reference to walking wave functions.

In words: the wavelength of a particle, λ, can be calculated by dividing what's known as Planck's constant, h, by the momentum of the particle, p. (Momentum, which is traditionally denoted as p in physics and engineering,[10] is just mass multiplied by velocity: $p = mv$.) That's it. We divide two numbers to find the wavelength. And yet this exceptionally simple, but hugely important, equation does not factor into any of the torrents of quantum woo out there. Let's see just why that's the case, starting with h, an exceptionally small number that nonetheless looms large in quantum physics.

Vanish into Small[11]

Despite its nondescript symbol, if there's one number that is the signature, multiplatinum, Rock and Roll Hall of Fame star value in quantum physics, it's Planck's constant. First introduced by the German physicist—and, coincidentally (or *was* it?) gifted musician—Max Planck in a truly groundbreaking, world-changing paper published in 1900, h is riven into the very fabric of quantum physics; it defines the quantization of light and matter. It's fair to say that Planck is responsible for putting the "quantum" into physics.

It's worth noting that Planck was certainly no antiestablishment rebel, working at the fringes of the discipline. Too often, particularly in the so-called "post-truth" culture that has attracted so much attention while I've been writing this book,[12] expertise and working knowledge are

---

[10] Despite being a physicist for over thirty years (if I include my undergraduate days), I am embarrassed to say that I have only now, when writing this book, looked into why momentum is denoted with a letter p. I knew that an m was out of the question (because mass had already snapped that letter up), but why p? It turns out that p is used because the Latin for impetus (a concept closely related to momentum) is *petere*.

[11] From Black Sabbath's "The Sign of the Southern Cross," of course.

[12] . . . to the extent that Oxford Dictionaries chose "post-truth" for Word of the Year 2016. A quote from the unique talent that is David St. Hubbins is particularly

dismissed or undervalued. In many online fora, experts (in just about any field) are regularly chastised for being too "establishment," for not being willing to be "open-minded,"[13] or, worse still, accused of being shills for some shadowy, conspiracist, neoliberal agenda. Yet even Einstein, vaunted as the lone outsider who shook the establishment to its core, was not entirely isolated from the physics community of the time. Of Planck, whose discovery Einstein himself said would require a rewriting of the laws of physics, their mutual contemporary Max Born—another quantum pioneer—had this to say:

[Planck] was, by nature, a conservative mind; he had nothing of the revolutionary and was thoroughly skeptical about speculations. Yet his belief in the compelling force of logical reasoning from facts was so strong that he did not flinch from announcing the most revolutionary idea which ever has shaken physics.[14]

Einstein, Planck, and the Danish physicist Niels Bohr disagreed to various extents on many aspects of quantum physics. Very public debates between Bohr and Einstein, in particular, played an integral role in the development of the field, raising important philosophical questions that echo to this day and have yet to be addressed in full. But they all agreed—as practically all physicists now agree—that Planck's treatment, right at the turn of the twentieth century, of how radiation is emitted by matter was the "Big Bang" for quantum theory; this is where it all kicked off.

You're actually very familiar with the physics problem that fascinated Planck and drove him, in what he later balefully described as

---

apposite here: "I believe virtually everything I read, and I think that is what makes me more of a selective human than someone who doesn't believe anything."

[13] As the great thinker Carl Sagan put it, "It pays to keep an open mind, but not so open your brains fall out."

[14] Born, M. (1948). "Max Karl Ernst Ludwig Planck. 1858–1947." *Obituary Notices of Fellows of the Royal Society.* 6 (17): 161–188. doi:10.1098/rsbm.1948.0024.

desperation, to introduce the quantum to physics. Or at least you are if you've ever switched on a lightbulb.[15] Planck was trying to understand the spectrum of light given out by an object that is heated up, rather like the filament in a standard incandescent bulb where an electric current is passed through the wire and it glows white-hot. It wasn't exactly a lightbulb filament that drew Planck's interest; it was instead (of course) an idealized version of the light emission problem. (He was a theoretical physicist. What do you expect?) But nonetheless, his query addressed the same white-hot issue at its core: Just how does matter emit light?[16]

That theoretical physics failed to provide illuminating insights into the origins of the emission spectrum was rather an embarrassment for physicists of the day.[17] Without going into the thorny mathematical detail, I'll sum up the problem (very, very loosely) by saying that for matter to emit light its atoms and electrons must get excited and jig about, and the rate at which they do this determines the wavelength of the light emitted. Alas, physics at the time was fairly clueless when it came to understanding how atoms and electrons vibrate and oscillate in

---

[15] I'm not the best at sticking to deadlines. To quote the masterful Douglas Adams, "I love deadlines. I love the whooshing sound they make as they go flying by." But if traditional incandescent lightbulbs have slipped from public memory by the time this book is published, I will have missed my deadline by a vastly, hugely, mind-bogglingly long stretch of time.

[16] A decade after Planck's work, this problem became known as the *ultraviolet catastrophe*. Wouldn't that be a great name for a metal band? I checked Google to see if it already had been snapped up, and it turns out there was a synth-pop band from Seattle in the early '80s who were indeed called the *Ultraviolet Catastrophe*. Why do I suspect that there may possibly have been a physicist in that band? Ah, well. Metal's loss is synth-pop's gain. By the way, it's very often the case in physics textbooks that Planck's pioneering work is described as being driven by the goal of finding a solution to the ultraviolet catastrophe. As David Darling describes in his fascinating history of the topic, the chronology doesn't support that interpretation. Tsk. A beautiful story killed by an ugly historical fact. See more on Darling's website: www.daviddarling.info/encyclopedia/Q/quantum_theory_origins.html.

[17] An embarrassment of *St. Anger* proportions. (Apologies, it's unfair to Metallica to single out that album. After all, there's a richness of embarrassment when it comes to their output from the *Black Album* onwards.)

solids. It didn't help that Planck was not a fan of the atomic picture of matter, having said back in 1882 that "[I]n spite of the great successes of the atomistic theory in the past, we will finally have to give it up and to decide in favor of the assumption of continuous matter." Much like his reluctant acceptance of the reality of quanta, however, he begrudgingly came round to the idea that matter was indeed made of atoms. Because that was where the data led him.

Unfortunately, these difficulties with understanding how matter emitted light were not a minor issue that could be swept under a corner of the fabric of reality and studiously avoided; it wasn't a question of some triviality that amounted to just dotting the i's and crossing the t's in the Big Book of Theoretical Physics. Matter vibrates and oscillates *all the time*. It's what atoms and molecules *do*. Ceaselessly. Even if we could reach the ultimate chilliness of absolute zero—and the laws of thermodynamics tell us we can't—they'd still vibrate.[18] Much like Marsha and her moshing friends at a Metallizer gig, they're restless (and often wild).

Planck was the first to realize that at the quantum level, vibrations can't have any old frequency—instead, they are quantized.[19] Just like those standing waves on a guitar string, only certain fixed frequencies are possible. That was how the vibrations of matter produced only the precise spectrum of light that so perplexed all other scientists at the time. This, to put it mildly, was not entirely expected. Indeed, Planck himself initially dismissed his quantization strategy as just a mathematical trick. He assumed that the quantized oscillations he required to make his theory fit with the experimental data were somehow virtual,

---

[18] We'll see just why in the next chapter. For certain. Or perhaps not.

[19] Before my physicist and chemist colleagues get cross because they think I'm cutting too many corners in the explanation here, I should stress that Planck wasn't specifically considering atoms. He spoke only of "oscillators" or resonators. Moreover, free atoms bouncing around in the form of a gas, as compared to those tethered to their neighbors in the form of molecules and solids, do not have this type of vibrational energy. There are, however, rather few examples of free single atoms of this type in nature. Atoms are generally gregarious and like to form bonds with their neighbors.

rather than real. It seemed a step too far—that "act of desperation" to which I referred earlier—to claim that atoms, or molecules, or *anything* could only vibrate at certain frequencies.

It turns out that those quantized oscillations are not only very real indeed, but there's a straightforward formula that describes how the energy is broken up into its tiny chunks. Here's the remarkable idea at the core of Planck's reasoning, an exquisitely simple mathematical equation that came as the result of a breathtakingly sophisticated insight into the fundamental nature of our universe. Energy comes in packets, in chunks, in *quanta*. And the energy, *E,* of one of those quanta is given by this formula:

$$E = hf$$

That equation, which many physicists consider to be every bit as elegant and beautiful as the rather more renowned $E = mc^2$, belies, in its remarkable simplicity, a seismic shift in our understanding. What it tells us is that energy at the (sub)atomic and molecular levels is *not* continuous. There's a fundamental limit to how small a change in energy can be, and it's given by multiplying the frequency of the oscillation, *f,* by that tiny constant that bears Planck's name. And Planck's constant really is very tiny indeed. We'll see just how quintessentially teensy those chunks of energy are very soon. But let's continue for a moment to focus on the quantization itself rather than get hung up on the size of the quanta.

The quantum nature of energy is rather like the difference between following the smooth, gently sloping highway to hell in the cartoon on the next page and descending instead the stairway to Hades. For the latter, your energy changes in discrete steps, rather than continuously. Either way, hell awaits, but the descent to the netherworld is rather more bumpy—more *discontinuous*—if you take the stairs.[20]

---

[20] But we metal fans all know that hell ain't a bad place to be in any case.

In the macroscopic world—even in our fictional hell-bound world—we can choose whether we take the stairs, or we can simply make the steps a different size when designing the staircase; it's not as if certain energies aren't accessible to us. But that's not the case in the quantum underworld: certain energies just *aren't* accessible.[21]

To make this point a little clearer, consider the following elevator pitch. Let's say Max Volumus takes the lift[22] up to his hotel room following a particularly energetic Metallizer gig earlier that night. The lift doesn't jump directly from the first to the second floor, and then jump

---

[21] "God runs electromagnetics on Monday, Wednesday, and Friday by the wave theory, and the devil runs it by quantum theory on Tuesday, Thursday, and Saturday." —William Lawrence Bragg (1890–1971). Bragg shared the 1915 Nobel Prize in Physics with his father, William Henry Bragg, for his work on using X-ray diffraction to determine the atomic structure of crystals. Bragg the younger also had more than a walk-on role in the discovery of the structure of DNA. But I'm getting ahead of myself (again).

[22] "If you travel to the States . . . they have a lot of different words than what we use. For instance: they say 'elevator,' we say 'lift'; they say 'drapes,' we say 'curtains'; they say 'president,' we say 'seriously deranged git.'" —Alexei Sayle (1952–)

again from the second to the third floor, etc. It doesn't teleport![23] It travels on a smooth, continuous transition between the floors. Similarly, when a circle pit broke out during the concert that evening, certain speeds weren't verboten; the moshing didn't discontinuously jump between slow and fast, with nothing in between. The speed gradually— or not so gradually—ramped up continuously.

Not so in the quantum world. Steps and jumps in energy are the signature of quantum behavior. Their discovery hugely unsettled physicists accustomed to dealing with the classical, continuous, predictably sloped physics of the everyday world. Max Born even went as far as to say that "no language which lends itself to visualizability can describe quantum jumps." Born's qualms notwithstanding, let's attempt to visualize a quantum jump. I warn you in advance that the following attempt at visualization will be misleading but, as with practically all analogies in physics, it's not so much a question of whether or not you're misled, it's a matter of the degree to which you're intellectually deceived. As an astronomer of my acquaintance put it, "[teaching] physics involves ever-decreasing circles of deception."[24] I'll try not to deceive you *too* much.

If we were to scale the elevator (and its rather discombobulated occupant) down to the quantum level, it would look a little like this:

---

[23] At least not at the time of writing this book. If you're reading this after teleporting home from a gig at CyberRockCity, it either means that this book has taken an *awfully* long time to reach the (virtual) shelves or we've very recently hit the technological singularity that certain futurists and pundits have been promising (for quite some time). And while we're on the subject of teleportation, the term "quantum teleportation" has got nothing to do with teleporting matter from one place to another (regardless of what some misinformed quantum pundits might tell you).

[24] See "Is Space Expanding?" from Peter Coles's blog *In the Dark*: telescoper .wordpress.com/2011/08/19/is-space-expanding/.

$$E_4 - E_3 = hf$$

$$E_3 - E_2 = hf$$

$$E_2 - E_1 = hf$$

To get the elevator from one floor to another in the quantum world, we need to inject a specific amount of energy, *E(nergy)* = *hf*. Unlike a classical lift, the quantum elevator will never get trapped between floors because it simply can't have an increase in energy that is smaller than *hf*. That is the minimum chunk of energy possible.

Similarly, electrons in atoms are constrained to "jump" from one level to the other without ever having an intermediate energy. This is the origin of the popular, and so often entirely misleading, "quantum leap forward" expression that has a habit of cropping up in just about any context, no matter how far removed from the *actual* quantum. You know the type of thing:

> **There has been a quantum leap forward in understanding how to plan for a successful retirement.**
>
> **Over the last five decades there has been a quantum leap forward in understanding how capital markets work.**
>
> **There has been a quantum leap forward in actions by international tribunals since the later 1990s.**
>
> —Genuine examples found via my thirty-second Google search

There are actually two misleading aspects here. The first is relatively easy to deal with and relates to scale. The *Oxford English Dictionary* definition of "quantum leap" is as follows: "noun. A sudden large increase or advance." But that's not quite right. A quantum jump/leap is certainly a discontinuous change and, as such, could be seen as a way of describing a major departure from the norm. This may help explain why the term has caught the public imagination. (The *Wow! Quantum!* imprimatur also certainly helps.) But given that you've made it this far in this book, one thing you'll certainly know by now is that the word "quantum" is not at all synonymous with a large anything.

It'll help to get things in proportion if we again consider quantum jumps in the context of that elevator. Let's scale our hotel elevator and its occupant back up to conventional dimensions. Assume that the elevator and Mr. Volumus together have a mass of 1,000 kilograms, and that the vertical distance between floors in the hotel is 4 meters. We can now very easily work out the change in potential energy as the elevator moves from the first floor to the second floor—we did precisely the same type of calculation back in Chapter 2. All we need to do is plug the numbers into the equation E_{pot} = *mgh*. In other words, we just multiply the three quantities *m, g,* and *h* together. We've already guessed that the mass, *m,* is 1,000 kg and the height to which the lift moves is 4 m. To keep the numbers crisp and round and even,[25] we're going to take the acceleration due to gravity, *g,* to be 10 meters per second per second.[26] The product *mgh* = 1,000 × 10 × 4 = 40,000 J. That's 40 kJ of potential energy.

Now let's compare that 40,000 joules of real-world elevator energy to the energy of a single quantum leap between floors for the lift's nanoscopic counterpart. Our equation for the quantum of energy is E = *hf*.

---

[25] Remember that, to a physicist, a cow is just a large sphere. And π is 3, give or take . . .

[26] Look back at Chapter 1 if you've forgotten where the extra "per second" comes from.

I've mentioned (just once or twice) that h, Planck's constant, is exceptionally small. Here's how small:

h = 0.00000000000000000000000000000000662607 joule-seconds.

In rather more scientific notation, that's 6.62607×10^{-34} joule-seconds. Don't worry about the joule-seconds units. Just why Planck's constant is expressed in units of joule-seconds could be the subject of not just a dedicated chapter but an entirely separate book.[27] The important thing to recognize is that when we multiply Planck's constant by a frequency, we'll end up with a quantity of energy. And because Planck's constant is so very small, that quantum of energy will be similarly exceptionally small.

So we know what h is, but what do we choose for the value of f, the oscillation frequency? There's a signature value here we can use. The atoms in a solid vibrate back and forth at a rate of about ten million million times a second, or expressed in a slightly more compact way (as physicists prefer[28]): 10^{13} Hz. That's an awful lot of vibrations per second. Not the tens, or hundreds, or thousands of Hz at which sound, speech, and music operate; nor the millions of hertz (megahertz, MHz) at which radio waves oscillate; nor, indeed, is it the blindingly fast billions of hertz (gigahertz, GHz) at which a computer processes

---

[27] Planck's constant is intimately related to an exceptionally important, but too often neglected (especially when it comes to teaching at high school and early undergraduate level), property in physics: the *action*. Very loosely speaking, the action tells us about the trajectories that a particle follows; it's about taking into account the history of an object's motion and the possible alternative pathways it could have followed. The principle of least action is an elegant—and many would claim, more powerful—alternative to Newton's laws that, unfortunately, does not get the attention it deserves outside of professional physicist circles. Richard Feynman was a major fan of the action concept, to the extent that the title of his PhD thesis was "Principles of Least Action in Quantum Mechanics."

[28] Following Dirac's example, physicists often prefer to be somewhat laconic in their communications. Not *all* physicists, mind you. We'll talk more about Paul Dirac in Chapter 7.

data. We've gone well past GHz speeds and are firmly in the terahertz regime—millions of millions of hertz.

Atoms are very small and very light—certainly as compared to an elevator—and that means they've got very high natural vibration frequencies. Every time a guitar string is struck, this fundamental relationship between size and natural/resonant frequency plays out. Shorten the length of a string and it will vibrate at a higher frequency: pluck an open string and it will sound a lower pitch than a fretted note. You're also probably familiar with the old singing wineglass demonstration.[29] Running a finger around the rim of the glass drives the glass to resonate at its preferred natural frequency. But fill the glass with wine or water, and the pitch of the note drops. Like many of those deceptively simple questions—"Why is the sky blue?"; "Why is glass transparent?"; "How the hell can my headphones get *that* tangled in only five minutes?"—the physics underpinning the change in the pitch of the note is remarkably sophisticated, as A. P. French described in substantial detail in a paper published in the *American Journal of Physics* back in the 1980s. French's mathematics are beyond even the "Maths of Metal" appendix, I'm afraid, but I can fortunately condense his analysis to a general principle: the more stuff there is, the lower the natural frequency. Or, to put it another way, the bigger things get, the heavier they sound.

---

[29] Much as for the pairing of quantum physics and metal, our favorite musical genre and wine generally tend not to be mentioned in the same breath. This book, however, is all about puncturing those heavy stereotypes (with just a little bit of physics thrown in), so I did some research into whether there are wine-metal connections that deserve more attention. And indeed there are. Far from the frustratingly widely held expectation that the average metal musician (or fan) is uncultured and uncouth in equal measure, metal has its fair share of connoisseurs of a good bottle of plonk. Tool's Maynard James Keenan currently owns Merkin Vineyards and the associated winery, Caduceus Cellars, while erstwhile Queensrÿche frontman Geoff Tate teamed up with the Washington State–based Three Rivers Winery back in 2008 to create Tate's "Insania." There's even a Spinal Tap–inspired wine: E11even Wines, from Andrew Murray Vineyards. I like to think of Nigel Tufnel sipping a glass of E11even while composing that sensitive, delicate, and harmonically subtle "Mach" trilogy he's been working on for a while.

TWANG FACTOR

If you haven't got either a guitar or wineglass at hand, grab a ruler,[30] hold it down on a table or bench so that it's partly overhanging the edge, and "twang" the free end. Vary the length that it is free to vibrate and twang again. This is not an entirely novel experiment, so I'm fairly confident that you're unsurprised by the result: the shorter the free length of the ruler, the higher the frequency of the twanged note. If you could somehow push this experiment to the point where the free length of the ruler was just one millimeter, and then one micrometer (a thousandth of a millimeter), and then one nanometer (a thousandth of a . . . you get the picture), the frequency of the twang would correspondingly get progressively higher. At the one nanometer scale, we're talking about a free length of the ruler just a few atoms across.

A very helpful way of visualizing atomic vibrations is to think of the chemical bonds between atoms of a molecule or a solid as springs. The stronger the chemical bond, the stiffer the spring. This is not just a "popsci" analogy that professional physicists and chemists might dismiss as too simplistic. There's a major area of computational chemistry known as molecular mechanics where this is exactly how chemical bonds are treated.

Stick a weight on the end of a spring and it will bob about at its natural frequency; change the weight or the stiffness of the spring, and the rate of bobbing will change accordingly. You may have encountered one of those "bobbing head" action figures? If not, here's a sketch of a bobblehead figure of some renown in the world of metal.

---

[30] If you're in need of a stylish ruler, I can recommend the University of Nottingham's very own Ruler of the Universe: http://www.scalerulers.co.uk/blog/the-ruler-of-the-universe-102/.

The rate at which Lemmy's head bobs back and forth is determined by the stiffness of the spring. (And just how heavy his head might be. But this is *Lemmy*. We know that it's going to be heavy.) If we change the stiffness of the spring (also called the spring constant), then the frequency at which Lemmy's head bobs back and forth will also change. There's a direct relationship between the resonant frequency and the stiffness of the spring. And this holds true all the way down to the atomic level: the chemical bonds are like nanoscopic springs.

We've finally reached the point where we can write down a value for a single quantum of vibrational energy. We just need to multiply Professor Planck's exceedingly small constant by the really quite big number representing the frequency of atomic vibrations, that 10^{13} Hz. As we're about to see, however, the smallness of Planck's constant nonetheless outweighs the largeness of the vibrational frequency. Let's do the maths.

Here's our simple formula again: $E = hf$. I'll first round up Planck's constant to just one decimal place—it will become clear in a second just why we don't need to worry about being exceptionally precise here. Plug the numbers into the equation and we have $E = 6.6 \times 10^{-34} \times 10^{13}$ joules. When we multiply powers of ten we add the exponents. This means that the energy of our vibrational quantum is $E = 6.6 \times 10^{-21}$ J. That's one thousandth of a millionth of a millionth of a millionth of a joule. Compare that invisibly small packet of energy with the 40,000 joules of the full-size elevator on the second floor. That's a factor of more than twenty-five orders of magnitude or *twenty-five* powers of ten: 10,000,000,000,000,000,000,000,000,000.

Of course, raw numbers generally aren't very helpful in letting us visualize relationships between different quantities—that's why scientists plot graphs. Here's a more visual representation of the difference in energy for the elevator when it's on the second, as compared to the first, floor. And right beside it is our vibrational quantum of energy on the same scale.

What do you mean that you can't see the quantum of energy in the cartoon?

Well, I *could* pretend it's there, for all that it matters—you're never going to be able to check, even if Pete, the illustrator, could draw the lines to scale in the first place. And he simply can't. No offense, Pete! No one can. No matter what state-of-the-art, hyper-high-definition graphic design program they may be using. If the size of a quantum of energy were to be shown in the diagram above in a similar way, and to scale, it would be represented by two lines with a gap of 10^{-26} meters between them.

10^{-26} meters.

That's a zero, then a decimal point, then twenty-five *more* zeroes before, finally, . . . a one. Forget about those old clichés where the size of something small is put in the context of the period at the end of this sentence. Those period analogies are worse than useless in this case; let's think on a more universal level.

The diameter of the observable universe is estimated to be 93 billion light-years. A light-year is the distance that light travels in a year. Light is rather nimble, so this is a very long distance indeed—nearly 10 trillion kilometers, 9.4×10^{15} meters. That means that 93 billion light-years corresponds to about 8.8×10^{26} meters.

The width of our observable universe is about as big as it gets. At the other end of the scale, let's take the size of a human for comparison. Average height is about 1.7 meters. This means that, give or take, the observable universe is a factor of "only" twenty-six orders of magnitude larger than you or me. Only.

That quantum of energy isn't just small, it's smallness on a *universal* scale. In relative terms, you are just as close in size to the vastness of the universe as the vibrational quantum is to the changes in energy you encounter every time you take an elevator, or walk down a street, or listen to even the most gentle and delicate of sounds (an Opeth acoustic interlude, say). When it comes to the type of energy levels encountered at a typical metal gig, the comparison with a single quantum is quite literally off the scale. The next time you hear someone opining on the influence of the quantum in our human lives, point out this massive disparity in energy scales.

A Perfect Circle

You might remember that I said there were two issues with how the term "quantum leap" is used. We've covered one of those—the question of scale. The second aspect is not quite so obvious but is no less important in understanding and appreciating the quantum world. The elevator analogy for energy levels is often used to explain quantum physics in concert with the popular planetary model of the atom, where electrons are constrained to certain "orbits" and an atom is the solar system writ small:

That sketch above is an entirely misleading picture of atomic structure[31]—because, as we've seen, electrons should be described in terms of probability waves rather than as point particles circulating in entirely orderly, "celestial" orbits—but it's nonetheless a conceptual model that even scientists often invoke. One of the reasons why the "electrons as orbiting planets" model still holds sway is that it captures the core idea of energy quantization and quantum jumps in a way that *can* be visualized.[32] It's badly wrong, and it's undeniably obsolete physics, but at least in some ways it's *helpfully* wrong.[33]

The idea of an electron having a specific, quantized energy is at the core of the planetary model. Just like the quantum elevator moving between different floors, the electron can't adopt any old orbit—only

---

[31] Even leaving aside the demonic nature of the electrons.

[32] The wave function, on the other hand, is not, in the language of quantum mechanics, an *observable*. It's something that, while it can be represented mathematically, does not have a physical "realization." That, as Born puts it, does not "lend itself to visualizability."

[33] Or as the talented science writer Jennifer Ouellette put it, "Don't Be Dissin' the Bohr Model": blogs.scientificamerican.com/cocktail-party-physics/dont-be-dissin-the-bohr-model/.

certain specific orbits, each with its own specific energy, are possible. It really is a case of the elevator jumping from floor to floor without ever existing between floors. And to jump from one orbit to another requires that the electron absorb a quantum of energy.

Those quantized energies mean there's a subtle link between Planck and Pentagram, between Bohr and Biohazard, between Einstein and Earache.[34] Just as the standing waves on a guitar string can't adopt any old frequency—only specific values are possible—the waves of electrons in an atom are similarly constrained. And as we saw for the guitar string, and the quantum corral, and the coffee cup, it's all a matter of confinement.

Bohr's model (sometimes alternatively called the Bohr-Rutherford model) of the atom assumes that the electrons circulate around the nucleus of the atom in circular orbits. The crucial point is that only those orbits that involve a whole number of wavelengths of the electron are permitted, as shown in the following sketch:

Does this sound familiar? Remember that for the guitar string, the physics of the situation means that there must be a node at either end (at the nut and bridge). The wave *must* have zero amplitude at those

---

[34] Having lived in Nottingham since 1994, it would be entirely remiss of me not to shoehorn in a reference to the influential Earache Records independent label. Founded in 1985 by Digby Pearson, with its first release in 1987, Earache is still based in Nottingham (although it now also has a New York office), and Mr. Pearson has led the company throughout its thirty-year history. It's been an exceptionally important label in metal over the years, particularly with regard to fostering the nascent grindcore and death metal scenes in the mid-1980s to early '90s. Napalm Death, Carcass, Godflesh, Nottingham's own Pitchshifter, The Dillinger Escape Plan, and the wonderfully monikered Lawnmower Deth are just a few of the metal bands that have featured on the Earache roster.

points because the string is fixed there. These boundary conditions define what wavelengths can be present, and that in turn means that only certain frequencies can exist. It's exactly the same idea for Bohr's model of the atom, except in this case the wave is confined to a circle. Imagine a circular guitar:[35]

The boundary condition here may be slightly different than for the string on the guitar, but it's the same principle we've encountered many times before. Only standing waves with the correct wavelength can form.[36] We wrote down simple formulae for the guitar string to

---

[35] Back in the 1960s, H. F. Meiners of Rensselaer Polytechnic Institute in Troy, New York, published a short paper describing this effect but with a violin, rather than a guitar, string. The diameter of the violin string was .008," which—as the guitarists among you will know—is the same thickness as the top E string in a "light-gauge" set of guitar strings. The paper, published in the *American Journal of Physics* 33, no. 14 (1965), was available online at the time of writing: aapt.scitation.org/doi/pdf/10.1119/1.1971061.

[36] It's important to highlight that we're talking about waves on the *perimeter* this time, not on the "surface" of the loop (which doesn't exist in this case anyway). When we discussed the quantum corral and the coffee cup, it was the standing wave formed within the circle that was key. Here, we're looking at a one-dimensional string looped round on itself. It's what's happening *on*, rather than inside, the circle that's now the issue. You might be wondering if there are conceptual parallels between these loops and string theory. There are indeed. But that's a matter for a whole other book.

determine the permitted wavelengths and frequencies, and we could similarly do some back-of-the-envelope algebra for the Bohr atom, but let's focus on the headlines.

Only a whole number of electron wavelengths is permitted for each orbit in the Bohr model. This in turn means that only orbits with a certain radius, r, are allowed because the values of r are constrained by the following rule:

$$n\lambda = 2\pi r$$

Let's tease that apart—it's not very different at all from the rule we saw for standing waves on a guitar string. The right hand of that equation is simply the familiar formula for the circumference of a circle, $2\pi r$. And just as for the standing waves on a guitar string, n is a positive integer: 1, 2, 3, 4, and so on . . . In other words, we have quantized wavelengths; the circumference of the circle must be made up of a whole number of these.

CRAZY IS AS CRAZY DOES

To be historically accurate, I should stress that this consideration of quantized electron wavelengths didn't come from Niels Bohr. He didn't interpret the orbits in his model in the context of the wave properties of matter, because that revolutionary counterintuitive idea was another decade or so down the line. It was Louis de Broglie who put Bohr's orbits in terms of electron waves. Bohr's model of the atom was already seen as rather "out there" at the time, and that was *before* electron waves were introduced to the mix. But Bohr was not one for dismissing apparently ludicrous ideas out of hand. Later in his career, he had this to say to his colleague Wolfgang Pauli: "We are all agreed that your theory is crazy. The question which divides us is whether it is crazy enough to have a chance of being correct. My own feeling is that it is not crazy enough."

This is not to say that every ludicrous theory is to be taken seriously. Like many physicists, I receive a steady trickle of email messages from "enlightened" individuals all across the world who have, they claim, solved all of the secrets of the universe through their own unique and revolutionary theory of everything. Without the burdensome irritation of any mathematics. Or quantitative predictions. Or the most rudimentary definitions of whatever the heck it is they're talking about. They, of course, have not attempted to have this work published because they know deep in their souls that their approach is simply too revolutionary for "the establishment." They are instead emailing it to academic physicists, helpfully written out in interesting FONTS, with lots of '90s vintage rotating gifs and flashing type, and a generous smattering of CAPITALS AND EXCLAMATION MARKS!!! Oh, and lots of disgruntlement about how Einstein was wrong and they can't believe that we've all been hoodwinked for so long.

Almost twenty years ago, as a service to the physics community, the mathematical physicist John Baez helpfully put together what he described as "a simple method for rating potentially revolutionary contributions to physics." Among the many gems in Baez's index, which has become even more valuable in the years since he first put it online due to the phenomenal growth of the internet as a platform for the most, errmm, *unique* ideas humanity has to offer, are the following: "10 points for beginning the description of your theory by saying how long you have been working on it. (10 more for emphasizing that you worked on your own)"; "10 points for each statement along the lines of 'I'm not good at math, but my theory is conceptually right, so all I need is for someone to express it in terms of equations'"; and, of course, "40 points for claiming that the 'scientific establishment' is engaged in a 'conspiracy' to prevent your work from gaining its well-deserved fame, or suchlike."[37]

In Bohr's model, the energy of an electron depends on the radius of its orbit. So what we have again is a sort of energy elevator. But it's

---

[37] I particularly like this "meta" twenty-point contribution to the crackpot score in Baez's index: "20 points for e-mailing me and complaining about the crackpot index (e.g., saying that it 'suppresses original thinkers' or saying that I misspelled 'Einstein' in item 8)."

a *quantum* elevator, so the electron jumps from one floor to another discontinuously—it's not "winched" up in a smooth fashion. How then does an electron make the transition from one floor to another? Just as for our vibrating atoms, it absorbs a quantum of energy, $E = hf$. Same formula, different context. This time we're not considering how atoms vibrate, we're down on an even smaller scale: we're shaking up the electrons within atoms. To take an electron to a higher level, we need to inject a quantum of energy. And one way of doing that is to shine some light on the problem.

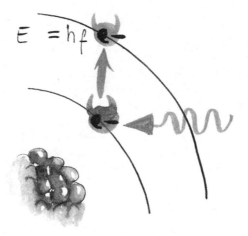

When Bohr came up with his model in 1913, Einstein had already shown— back in 1905, five years after Planck rocked the physics world with his quantum insight—that light also comes in quanta. In a paper with the snappy title of "On an Heuristic Point of View Concerning the Creation and Conversion of Light"[38] he wrote:

> *In accordance with the assumption to be considered here, the energy of a light ray spreading out from a point source is not continuously distributed*

---

[38] Or, in the original German (published in *Annalen der Physik* 17 (1905): 132–48), *"Über einen die Erzeugung und Verwandlung des Lichtes betreffenden heuristischen Gesichtspunkt."* They didn't really have the equivalent of clickbait back then. Bliss it was in that dawn to be alive!

over an increasing space but consists of a finite number of energy quanta
which are localized at points in space, which move without dividing, and
which can only be produced and absorbed as complete units.

It's only the last part of Einstein's sentence that really need concern us. Light is made up of "energy quanta . . . which can only be produced and absorbed as complete units." Once again we have those packets, those chunks, those *quanta* of energy. An electron can absorb an energy "pill" of light, but only as a unit. It's not a question of absorbing half the amount of energy and making half a transition: it's an all or nothing situation. Discrete, not continuous. Lumpy, not smooth.

It's a little known fact outside of physics circles, but Einstein wasn't awarded the Nobel Prize for his work on the theory of relativity. He instead won the accolade for precisely what we've been discussing: insights into the nature of light and how it interacts with the atomic building blocks of matter. Specifically, for his explanation of a fascinating phenomenon known as the photoelectric effect (or photoemission). Physicists use photoemission all the time in our research (in common with many, many other scientists across the world). It's an easy phenomenon to describe. (Rather less easy to explain. Hence Einstein.) Shine light on a sample, and electrons are excited from one energy level to another. With light of high enough energy—and remember, the higher the frequency the more energetic those quanta will be—electrons can even escape from their parent atom and fly free to the outside world.

But that's enough of Einstein's quantum achievements for now. There's an entire chapter dedicated to the physics of light (in, of course, a suitably metal context) coming up. Let's get back to the nanoscopic solar system of the Bohr atom—and my second objection to the term "quantum leap."

The concept of a quantum jump is very often couched in terms of a physical leap for an electron, like a tiny planet kicked from one orbit into another. And indeed, that was Bohr's model. But that's really not how it works down there in the nanoworld. It's at this point that Bohr's model starts to become rather less helpfully wrong and can set up some serious misconceptions. Let's try to dispel some of those by returning to the subject of Battery.

Oops, sorry. I meant *batteries*.

Returning the Reaction

There's a huge difference between a leap in energy and a physical jump. Yes, in the real world, changes in energy are very often associated with physical motion.

But often they're not. As just one example, let's consider what happened when Marsha the Mosher charged her phone before taking it with her to the gig above. (Allowing her to illuminate those nearby with that soothing screen glow during the most dramatic interlude in Metallizer's set, when the lights were low and the lighting technician was working so hard to create the appropriate mood.) First, and this may come as a surprise, but when Marsha charged her phone, she didn't actually *charge* her phone; she didn't directly add a dollop of positive or negative charge to it. A battery has zero net charge. Confusing, I know, but it's another example of where everyday language and colloquialisms not only don't quite capture the underlying physics but can be somewhat misleading.

A battery isn't a box stuffed full of charge that all comes pouring out when it's connected to power a phone or a laptop or a guitar FX pedal. It's much more subtle than that. When Marsha charged her phone, it wasn't as if she was doing the electronic equivalent of filling up a bucket of water. I know this is how battery charging is often explained, and I've been guilty of sometimes lapsing into those water-bucket analogies myself, but, much like an electron undertaking a quantum leap in Bohr's atom, we're going to a deeper level in the hierarchy of those circles of deception.

Think of the battery not as a tank full of charge but rather as a source of chemical energy that can move charges around. Chemistry is all about understanding and controlling how electrons are exchanged, shared, and put to work in reactions,[39] so perhaps it shouldn't come as a surprise to hear that, ultimately, chemical reactions are at the root of our mobile communications.[40] Batteries exploit a particular type of chemical reaction to provide the electrical energy—the "oomph"— required to drive charge around an electrical circuit.

Mobile phone (and laptop) batteries are, at the time of this writing, of the lithium-ion variety.[41] You'll be relieved to hear that we don't need to get into the technicalities of the technology in order to get a reasonable working understanding of just how these ubiquitous batteries provide what might best be called, with all due credit to the Young

---

[39] Yes, my fellow chemists, I realize that there's a little more to chemistry than this. But I hope you'll forgive the simplification for the purposes of clarity. *"Um, hang on. Did he just say "fellow chemists"? I thought he was a physicist."* I am indeed formally a physicist. But a particularly exciting aspect of nanoscience is that it blurs the boundaries between traditional disciplines. The research our group does could equally well be described as physical chemistry or chemical physics. We need a lot more of this blurring of the boundaries that happens in nanoscience. Too often science is carried out in "silos," which are by and large a historical accident in terms of how different fields have evolved.

[40] Or, as expressed somewhat more lyrically by that Torontonian trio: *"Signal transmitted. Message received. Reaction making contact."* (From *Signals*, Anthem, 1982.)

[41] I suspect, and hope, that this sentence will be quaintly out of date very soon after this book is published. There is a great deal of worldwide research into higher efficiency battery technology. There was a flurry of activity in 2016. It will be fascinating to see how much of this fundamental battery research makes it out of the lab and into the big, bad world.

brothers and Mr. Bon Scott, our powerage. A battery is made up of two electrodes—the positive and negative terminals—embedded in an electrolyte. Without going into the chemical minutiae, lithium ions (i.e., lithium atoms that have lost an electron so that they have a net positive charge) travel through the electrolyte from one electrode to the other, both when the battery is being "charged" and "discharged."

When Marsha's phone is charging, those lithium ions travel from the positive terminal to its negative counterpart, but when she's busy snapping as many selfies as possible during the Metallizer gig, the flow of charge is reversed. Plugging in her phone to charge it effectively resets the chemical reaction responsible for the motion of the lithium ions. The net result of this charge flow within the battery, which can only happen when its terminals are connected to the outside world, is that it creates a difference in the energy an electron has at the positive electrode compared to that it has when it's at the negative electrode.

In other words, the chemical reaction happening inside the battery establishes an energy *gradient* for the electrons. It's conceptually similar to the energy gradient the crowd-surfing mosher in the cartoon below experiences.

When he's held up by the crowd, the mosher's gravitational potential energy is higher. When we make the supporting crowd vanish—and because this is another *gedankenexperiment* we can do just that—then that difference in gravitational potential energy means that he will move due to the gradient in the potential. Net result? A fairly painful encounter with the floor of the concert hall.

Similarly, there's a *gradient* in energy that drives both the transfer of lithium ions inside the battery *and* the flow of electrons from one terminal to another in the surrounding circuit. This powers Marsha's phone and allows all those wonderfully flattering selfies to be captured for posterity. For the battery, it's not gravitational energy that's of importance[42] but rather a combination of electrical and chemical energy known as the electrochemical potential.

These gradients in energy (of whatever type) are key not just to how batteries work (and to the misadventures of moshers), but to virtually every aspect of the physics of our universe. A gradient in potential energy is a *force*.[43] As just two examples: for the battery it's known as the *electromotive* force, while the gradient in the gravitational potential energy experienced by our falling mosher goes by a rather better known name: the force of gravity.

What we shouldn't do, however, is confuse and conflate these different concepts. It's so easy for this to happen, however, because in order to explain tricky concepts in physics, we'll always reach for real-world examples, even though those examples might be doing more harm than good in some cases. And that brings me back to those quantum leaps we were discussing.

As I hope our battery example clearly highlights, there's a difference between a jump in energy and a physical jump. As your phone is being charged, it's not like its height above the floor increases; a jump in electrical or chemical energy does not lead to your phone levitating in midair. Fairly obvious, right? But this conflation of a physical change

---

[42] Gravity plays no role at all in the reaction. Take your phone to the moon, or to the depths of space, and its battery will work just as well.

[43] Yes, fellow physicists, I know, I know. Technically I have to multiply the gradient in potential energy by -1 to get the force. Let it go. It's the bigger picture that's important here.

in position with an increase in energy is exactly what too often happens when quantum jumps are being explained. Even in the macroscopic world of malls and metal, a change in energy does not necessarily mean that an object shifts its position. In the quantum world this distinction between energy and position is even more important, and yet it's not always appreciated (even by physics students).

The energy of a particle and the wave function that describes that particle are two very distinct things. For one thing, the energy is just a *number*—it's a certain amount of joules. Granted, it's an exceptionally, mind-bogglingly tiny amount of joules, but it's nonetheless just a number. The wave function, on the other hand, is a *function*—it's a way of describing a quantity that varies in space and/or time. It's not just one specific value. Moreover, an electron's energy may change, but it could in principle still be found at the same position afterward. Think back to Chapter 5. The electrons trapped in the nanostring were associated with wave functions that were, for all intents and purposes,[44] identical to the standing waves formed on a guitar string. However, just as each harmonic on a guitar string is associated with a different frequency, so each of those harmonics for an electron is associated with a different energy level as well.

This is quite similar to Bohr's model in that there are discrete energy levels. But there's one hugely important difference. Bohr thought that an electron was a nanoscopic planet: an orbiting point particle. As we saw earlier in this chapter, it was de Broglie who made the intellectual (quantum) leap to realize that if light—which everyone knew behaved like a wave—could be treated as a stream of discrete particles of energy called photons (as Einstein had shown), then why the heck couldn't it work the other way round? Why couldn't matter—i.e., particles like electrons, protons, and neutrons—behave like a wave? Well, they could,

---

[44] There are differences but, surprisingly, they're rather minor. As we discussed in the previous chapter, the one-dimensional chains of atoms in this case behave very much like the ideal "particle in a box" of a first-year quantum physics university course.

and they can. I'll remind you of de Broglie's very simple formula for the wavelength, λ, of a particle:

$$\lambda = h/p$$

Remember that in the formula, h represents Planck's constant, and p is the momentum of the particle. Therefore, there's a direct link between wavelength and momentum. That in turn means that the kinetic energy of a particle and its wavelength are intimately connected. Two lines of maths can show this (but feel free to skip forward past this, if you like, to get to the result on the next page).

We saw back in Chapter 2 that kinetic energy is given by this formula:

$$E_{kin} = \tfrac{1}{2}\, mv^2$$

where m represents the mass of the particle and v is its speed. Remember that momentum, p, is just the product of the mass and velocity,[45] $p = mv$. If we take the equation for the kinetic energy of a particle and plug in that momentum formula, here's what we get:

$$E_{kin} = \frac{p^2}{2m}$$

We already know from de Broglie that the wavelength and momentum of the particle are reciprocally related to each other. In other words, as the wavelength decreases, the momentum increases. And if the momentum increases, so too does the kinetic energy. (In fact, it rises in line with the square of the momentum, as that equation above shows.)

---

[45] We'll gloss over a really important aspect of momentum here—it's a vector quantity. This means it has a direction as well as a magnitude. For example, if Metallizer's tour bus is traveling southbound at 80 mph on the M1 motorway in England, it has a different momentum than if it's traveling northwards at the same speed. But it has the same *amount* of momentum. The direction of the momentum is exceptionally important in many aspects of physics, but for de Broglie's formula we only need worry about the magnitude of the momentum, not which way it's directed.

If you'll allow me just one more bit of maths, let's combine de Broglie's equation with that formula for kinetic energy just so we can see the relationship between the kinetic energy and the wavelength:

$$E_{kin} = \frac{h^2}{2m\,\lambda^2}$$

So, as the wavelength goes up, the kinetic energy goes down. (And it goes down quite quickly because it's proportional to not just $1/\lambda$ but to $1/\lambda^2$.)

───────────────────────

If you skipped the maths, welcome back. The central message that emerges when we bring those equations together is that the kinetic energy of a particle increases when its wavelength *decreases*. This means that higher harmonics are associated with higher energies. Let's revisit just the first and second harmonics for an electron in a nanostring, as in the sketch at the top of the next page. I've explained that from those harmonics—the wave functions—we can work out the probability of finding an electron in a certain region of space. But I haven't yet told you how.

You might think that some horrendously complicated mathematical abomination is about to raise its ugly head. Rest easy. It turns out that all we have to do to get a probability map for the electron is square the wave function[46]—that is, multiply it by itself. If we take a sine function and multiply it by itself we get a sine squared function (also known as a $\sin^2$ function). For the second harmonic, it looks something like this:

───────────────────────

[46] In general, it's a special type of squaring that I cover (where else?) in the appendix. (You didn't think that it was going to be as straightforward as just a "vanilla" squaring, did you? Nature isn't that kind to us.) But for the case of the electron in a nanostring, we're lucky in that common or garden-variety squaring does indeed work. Quite *why* it works is an entirely different matter that is definitely outside the scope of this book!

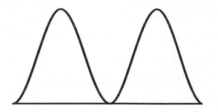

For each harmonic, the sine squared function tells us the probability of finding an electron in a certain region of the one-dimensional "box" (i.e., the nanostring). And the way we work out this probability is by taking the area under the curve. The electron has to be *somewhere* in the box, so if we take the entire area under the curve, this must be equal to 1. If we want to know the probability of finding an electron on the right-hand side of the box then we need to evaluate the area under the right-hand side of the curve. For the first harmonic, it looks like this:

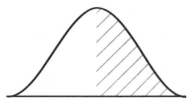

And for the second harmonic, it looks like this:

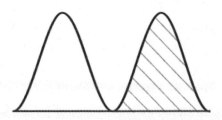

Now here's the important bit regarding the distinction between energy and position. Let's consider the first harmonic. The electron has a certain energy when it's in this state. Before we make a measurement, we don't know where the electron is going to be, but we *do* know how the probability of finding the electron varies: it's given by that sine squared curve. For the first harmonic, the electron is most likely to be found in the center of the nanostring.

If we inject a quantum of energy (like a photon of light), we can excite that electron to a different energy state: the second harmonic. But this is not a physical jump. The electron hasn't jumped out of the box or moved to a higher "orbit." It's still in there, cozy and confined. It's just that the probability of where we'll find it has changed. Compare and contrast:

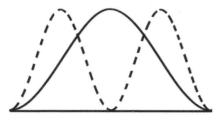

For the first harmonic (the solid line above), the electron is most likely to be found at the center of the box, while for the second harmonic (the dashed line), the electron is *least* likely to be found there. That's because, just as was the case for those opening notes of "Red Barchetta" we looked at back at the end of Chapter 4, there's a node at the center of the "string" for the second harmonic. At that point the sine function is zero. Nada. Zilch. And zilch squared is still zilch. So the center of the box is the very last place we'd expect to find the electron.

However, we still have a strong probability of finding the electron in one half of the box or the other, just as we did for the first harmonic. We can find the electron in precisely the same region of space, even though it has a very different energy in each case. If we have a mental picture of a quantum leap as a physical jump of a particle, then this doesn't make much sense, but in the context of Marsha's fully charged phone, it's hardly surprising at all. Her phone, despite having a higher level of chemical energy when it was charged, didn't change its physical position. Similarly, the quantum leap the electron experiences is a change in its energy. This means that the probability of finding it in different regions of space changes, but there is no physical "jump." Energy and position are very different quantities.

Although de Broglie was not at all comfortable with what turned out to be this exceptionally successful probability-wave interpretation of quantum mechanics—preferring to think in terms of a real, tangible wave—his formula remains entirely valid and is of exceptional importance in all scientific disciplines.

I'm going to close this chapter by giving you the key, and only, argument you need to cut quantum woo down to size. We've considered the wavelength of a tiny particle like an electron. But what's the wavelength of a human-scale object?

There's no more appropriate human-scale object to consider than, um, a human. We'll base our calculations on a giant of the metal world: the late, great Ronnie James Dio. His massively influential, commanding voice notwithstanding, Dio was somewhat diminutive in stature, a mere 1.62 meters (5' 4") tall. He was also someone who, although producing some of the weightiest music in the metal genre, was not physically the heaviest of men. Although I don't know precisely what Dio's weight might have been, I'm going to make a reasonable estimate and suggest that his mass was 60 kilograms (roughly 132 pounds).

Dio didn't move around a huge amount onstage but let's say that, between bouts of flashing his trademark sign of the horns,[47] he walked at a speed of 1 meter per second (3.6 km/hour).

---

[47] There's considerable debate in metal circles as to who first introduced the now ubiquitous sign of the horns to the genre. Dio, of course, was single-handedly responsible for popularizing "the horns" from shortly after he joined Black Sabbath in 1979 and subsequently throughout his solo career. In a significant number of interviews, RJD said that he adopted the horns because he wanted a signature stage gesture that would set him apart from Ozzy Osbourne, who frequently flashed the two-fingered peace sign during his gigs with Sabbath. For Dio, the horns originated with his grandmother, who used them to ward off the "evil eye." It's a hotly disputed issue, however. Never one to miss a branding opportunity, Gene Simmons of Kiss fame has claimed that he was using the horns sign long before Dio. But then Gene has always been a man of a thousand faces—my money's on Dio.

Let's plug the numbers into de Broglie's formula. We first need to work out Dio's momentum. His mass, m, is 60 kg, and his velocity, v, is 1 m/s. Although, as I said in an earlier footnote, momentum is a vector quantity (it has a direction—if Ronnie is walking left to right, he has a different momentum vector than if he's walking right to left), we only care about the *amount* of momentum for de Broglie's equation. So Ronnie has a momentum of 60 kg m/s. Let's plug that value, along with Planck's constant, into de Broglie's equation: $\lambda = h/p$. You know that Planck's constant is really very small indeed, so you can probably see where this is going already.

It turns out that Ronnie has a de Broglie wavelength of roughly 1×10^{-33} meters. Once again, this is so inconceivably small and so exceptionally tiny compared to anything in our everyday experience, it's not possible for our human cognitive system to grasp it. It's entirely beyond our ken. (And our Ron, for that matter.) Compared to a quantum particle, we are, in the words of that Strapping Young Lad known as Devin Townsend, *Heavy as a Really Heavy Thing*. Given that a nucleus is a femtometer across, 10^{-15} m, even if we were to scale down by the same factor again—to first shrink a human to the size of a nucleus and then shrink that nuclear human by the same amount—that's still a thousand times larger than Ronnie's de Broglie wavelength![48] We're not just giants when compared to the close-to-infinitesimal subatomic world, we're gargantuan, heavyweight behemoths who exist on scales that decidedly are classical, not quantum, in their scope. Of that we can be certain.

_____

[48] What's more, I'm assuming that we're just a single massive particle: an electron scaled up to Ronnie's size and mass. We're obviously slightly more complicated than just a single, coherent particle: we're a collection of molecules, with all their associated atomic and electronic structure, interacting in a head-spinning multitude of different ways. And each of those interactions will scatter the quantum waves.

Chapter 7

UNCERTAINTY BLURS THE VISION

Heisenberg and Schrödinger get pulled over for speeding.
The cop asks Heisenberg, "Do you know how fast you were going?"
Heisenberg replies, "No, but we know exactly where we are!"
The officer looks at him confused and says,
"You were going exactly 108 miles per hour!"
Heisenberg throws his arms up and cries,
"Great! Now we're lost!"

—Old physicist's joke (Anon)

Nobel laureate who helped discover quantum mechanics.
Name immortalized in pop culture as a meth dealer.

—Heisenberg internet meme

Nominated for the Nobel Prize by Albert Einstein and later a recipient of the coveted honor, Werner Heisenberg was indubitably one of the pioneers of quantum physics. Unfortunately, his insights and legacy have been rather scrambled by not only the *Breaking Bad*[1] connection but by the frequent mangling over the years of the meaning of his famous uncertainty principle. Truth be told, even Heisenberg himself was rather uncertain about the implications of his principle. Robert Crease, historian and philosopher of science at Stony Brook University, New York, had this to say on the subject of Heisenberg's uncertainty:[2]

> *Heisenberg thought the attempt to construct a visualizable solution for quantum mechanics might lead to trouble, and so he advised paying attention only to the mathematics. Michael Frayn captures this side of Heisenberg well in his play* Copenhagen. *When the Bohr character charges that Heisenberg doesn't pay attention to the sense of what he's doing so long as the mathematics works out, the Heisenberg character indignantly responds, "Mathematics is sense. That's what sense is."*

Crease goes on to point out in that interview just how important the uncertainty principle has been in popularizing quantum physics and bringing its weirdness to the masses: "The uncertainty principle's appearance in 1927 changed [the public perception of quantum physics]. Suddenly, quantum mechanics was not just another scientific

---

[1] If you're not familiar with the television series *Breaking Bad*, then I'm afraid I'll have to ask you to do some research as homework. Even a brief explanation of the plot would take up the remainder of this book.

[2] Quote taken from a great interview with Crease by the talented science writer Philip Ball: philipball.blogspot.co.uk/2014/10/the-moment-of-uncertainty.html. Ball's blog, *homunculus*, is enthusiastically recommended.

theory—it showed that the quantum world works very differently from the everyday world."

The (in)famous ethereal aspects of the quantum world, often perceived to arise from the uncertainty of Heisenberg's principle, play heavily into its mythology. It's ghostly down there; it's all about probabilities; nothing's solid; everything's uncertain.

Except it's not.

We've already seen that it's entirely possible to move individual atoms into well-defined patterns: the quantum corral and the atomic "strings" that we saw in Chapter 5 are remarkable feats of atomic-scale engineering. Each atom is, in essence, a building block that can be pushed across the surface to form a regular and stable pattern, just as we can connect Lego blocks to form different structures.[3] As long as the vacuum pumps keep working (so that contaminant gases can't gather on the surface and react with the atoms there) and the temperature is kept low (so that the atoms don't start hopping around as they start to warm up), the atomic corral and the gold chains we looked at in the previous chapters will remain solid as a rock. And the electron standing wave formed within the corral or chain will behave in an entirely predictable manner: we'll always see the same wave pattern when we take a measurement.

But we don't need to go as far as the sophisticated nanoengineering of the quantum corral or the gold atomic "string" to see this. Highly regular and entirely predictable arrangements of atoms—and, therefore, electrons—are all around us in the form of crystals. I'm writing this in December in Nottingham, England, where the temperature got to a couple of degrees below zero last night and there are plenty of ice crystals to be seen outside. As in any crystal, the molecules in ice are

---

[3] I recently discovered that there's a metal version of *The Lego Movie* theme tune, "Everything Is Awesome": musicfeeds.com.au/news/lego-goes-heavy-metal-crushing-cover-everything-awesome/.

packed into a uniform arrangement.[4] Ice crystals are particularly fascinating because there are, at last count, sixteen different possible packing
arrangements for the H_2O molecules. The form of ice that I see when I
look out my window on this rather chilly morning, and the form of ice
that is of key importance for life on Earth because it's found all across
our planet, is known as ice 1_h ("ice one h"). It's made up of sheets of
water molecules arranged in a hexagonal pattern. Entirely predictable.
So predictable, in fact, that we can depict the arrangement of molecules
with simple diagrams made up of balls (representing the atoms comprising the molecules) and sticks (representing the bonds between the
atoms).

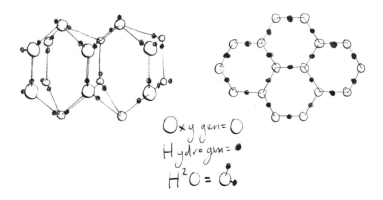

Regular. Repeatable. Predictable.

Ice is actually a bit complicated in terms of its crystal structure
(despite water being a rather simple molecule), so I'm going to use
another crystal as an example to drive home this point about the pre-
dictability and regularity that so often exists at the atomic and molecular
level. A few years back, our research group at the University of Nottingham showed that it was possible to flip a switch made of just two atoms.
And not only could we do that once, we could do it time and time

---

[4] Some arrangements are much more regular than others, however. There are
forms of ice that have an amorphous structure—the packing of the molecules is
much less regular. We'll just consider the more orderly forms of ice.

again. The atoms weren't ghostlike, wavy entities—they indeed acted much like teensy-weensy billiard balls. We could locate and relocate an individual atom, and flip its position. And we could do this with not only atomic precision, but with an accuracy considerably better than the diameter of the atom—we could locate individual chemical bonds.

Nature in this instance was kind to us. On the particular silicon surface we used,[5] the atoms pair up to form dimers. (We saw dimers earlier when we considered gold chains. This is exactly the same idea, except that these are silicon rather than gold atoms pairing up.) This leads to rows of dimers running along the surface, as shown in the sketch on the next page, where we're looking down on the surface from above. The silicon atoms pair up because it means they can form an additional bond. As mentioned previously, atoms are generally rather gregarious and love to bond with their neighbors. The bonding lowers their energy, and, importantly, atoms *want* to be at their lowest energy—at home on the couch, not leaping about in a circle pit. (As you may recall me saying in Chapter 2, there are few questions in physics that can't be answered by considering the lowest energy state of a system and how it might get there. In actuality, the reason that everything tends toward lower energy states has to do with statistics and thermodynamics and not anthropomorphic traits like laziness, but for now, for our purposes: atoms = gregarious and lazy.) The silicon atoms can lower their energy still further by "buckling": one atom moves down into the surface, the other moves up. And they do this in a zigzag manner so that neighboring dimers are buckled in different directions. (This alternate buckling arrangement places less strain on the atoms underneath.)

---

[5] Depending on the direction in which we cut through a crystal, we'll expose a surface with a particular, characteristic arrangement of atoms. Predicting just how those atoms will arrange at the surface, however, is notoriously tricky. Wolfgang Pauli, a huge name in twentieth-century physics (and, as mentioned in the last chapter, a subject of Bohr's commentary), said, "God made solids, but surfaces are the work of the devil." The invention of the scanning probe microscope was therefore a godsend to physicists, chemists, and materials scientists: those intricate arrangements of atoms at surfaces could finally be seen.

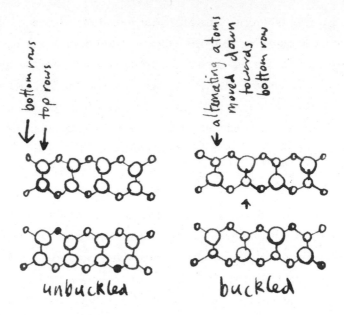

unbuckled   buckled

Could the tip of a scanning probe microscope be used to convert a dimer from one buckled configuration to another? In other words, is it possible to flip a two-atom switch by pressing down on one end? It turns out that the answer to this question is "not quite." We can't push down on one end because, as I said, atoms love to bond. If we push on one end of the atomic switch with the scanning probe, what happens is that a chemical bond forms between the atom at the very end of the probe and its counterpart at the surface. We can indeed push down, but when we try to lift our "finger" off the switch, it sticks to our fingertip and we end up pulling the atom back to its original position.

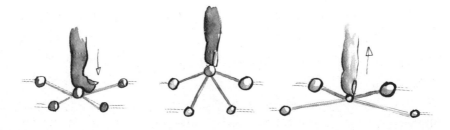

So we need to adopt a different strategy to flip the switch. If pushing doesn't work, how about pulling? Since atoms are so clingy, why don't we instead target the lower atom of the dimer, grab it, and pull it up? That's the proposed strategy. But does it work? Well, given that I've spent quite some time setting the scene, you can probably guess the answer. Here's the experimental evidence that it is indeed possible to flip an atomic switch:

Target atom

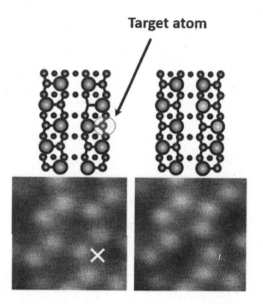

🎥 If you'd like to see an atomic switch flip in real time, we've got it on video here: www.youtube.com/watch?v=UBmBMmuUBMk.

The eagle-eyed among you will note that not just one, but two, dimers flipped. This coupling tells us a lot about what might be called the "elasticity" of the surface. By flipping a dimer, we change the strain on its neighboring atoms, and they in turn move to readjust their energy. If we had a perfect, defect-free surface, flipping one dimer would trigger a chain reaction and the others would flip to compensate. But not only are we competing against other defects at the surface, there's also the question of how the energy released by flipping a dimer is transferred to the other dimers. Ideally, all of that energy would transfer along the surface but in reality some of it leaks into the bulk of the sample, reducing the possibility of a long-range "cascade."

I should stress that, unlike the approach the IBM team used to construct the quantum corral, we didn't rely on the tunnel current flowing between the tip and sample to manipulate the atomic switch. Instead, we'd bring the probe *really* close to the sample so that a chemical bond formed. To build up an image of the atoms, instead of measuring a flow of electrons we swept the probe back and forth across the surface, measuring the variation in the strength of that chemical bond.[6] This breed of SPM is called, perhaps unsurprisingly, an atomic force microscope and was again, like the STM, invented at the IBM research labs in Zurich.[7]

But how do we measure the strength of a single chemical bond? It turns out that the best way to do this brings us full circle (pit) to the concept of resonance covered in previous chapters. We, and many research groups across the world who manipulate single atoms and molecules on a daily basis, use a tiny tuning fork to detect those chemical forces. A scanning probe tip is formed by first gluing a very thin piece of wire (a few hundredths of a millimeter across) onto one prong of the fork.[8] The wire is then etched to form a sharp point—this is our probe.

---

[6] I'm skipping over quite a bit of the detail here, of course. Technically, it's not the force of the chemical bond we measure. Rather, it's the gradient of that force, i.e., how quickly (or slowly) the force changes as we move the tip and sample closer or farther apart. We subsequently convert that rate of change into a force between the two atoms (the atom terminating the probe and the atom at the surface).

[7] Binnig, G., Quate, C. F., and Gerber, Ch. "Atomic Force Microscope." *Physical Review Letters* 56, no. 9 (1986): 930–33.

[8] Yes, gluing a wire that is less than a tenth of a millimeter in diameter onto a tuning fork whose prong is about half a millimeter across is indeed a *heck* of a lot of fun. Filipe Junqueira, one of the PhD researchers in the lab who had the enviable task of doing this most recently, has had hours of enjoyment with this particular activity.

The tuning fork, just like a guitar string, has its own natural resonances (although the relationship between the frequencies of the various harmonics is quite a bit more complicated than the string version). And just as for the guitar string, those resonant frequencies tell us a great deal about the forces acting on the tuning fork. We can use changes in the resonant frequency of the tuning fork as the probe scans across an atom or a molecule to determine the strength of the force between the atom right at the apex of the tip and its counterpart below. Although this might seem like esoteric physics in the context of single-atom imaging and manipulation, it's very much a real-world, everyday effect. And, as ever, there's a metal connection.

Just as we discussed in the context of the bobbleheaded Lemmy in the previous chapter, we can again think of the bond between the end of the probe and our target atom or molecule on the surface underneath as being a miniature spring. A little like this:

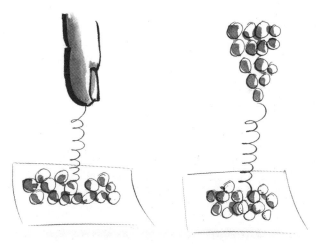

As the probe moves back and forth across the surface, from spring to spring, the stiffness varies. And the resonant frequency of the tuning fork changes in concert with the spring: the fork vibrates slightly faster or slightly slower depending on the "springiness" of the chemical bond between the probe and the atom below. In order to build up an image of the atom, the resonant frequency of the fork is measured at each point of the scan (each pixel in the image) and mapped onto a color scale. As for the scanning tunneling microscope we discussed a while back, we can choose whatever color scale we like. Atoms can appear in a range of colors or shades (and, depending on the state of the tip, shapes): crimson red, hellish green, deep purple, or (back in) black.

Given that we're using a tuning fork to make the measurements, can we *listen* to that fork? Can we *hear* atoms bond? Well, in a sense we can. Sort of. The fork vibration is exceptionally small—in many modern experiments it oscillates up and down by much less than the diameter of an atom—so there's no way we'll hear anything from the fork itself; not enough air molecules are driven back and forth by that ultratiny oscillation. But the fork generates a small electrical signal. That signal is amplified and then usually converted into a visual image by mapping the frequency to a color, but we can feed the fork's output into an amplifier—a Marshall amplifier, of course.

We did this for a *Sixty Symbols* video[9] a number of years back, where the signal from the tuning fork was fed out to a Marshall practice amp. (A 100-watt stack unfortunately wasn't available at the time.) Here's what the signal looked like as the tuning fork (and, equally importantly, the probe glued to its end) scanned back and forth over the surface:

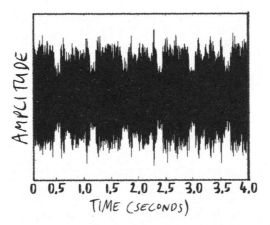

---

[9] *Sixty Symbols* (www.sixtysymbols.com) is a video series produced by video journalist Brady Haran for the University of Nottingham. View "The Sound of Atoms Bonding" at: www.youtube.com/watch?v=Ehw8PTA4QkE.

There are regular "dips" in the signal every 0.6 seconds or so. This periodicity arises from the regularity of the arrangement of the atoms at the surface: the probe encounters the same pattern (and thus the same changes in force) over and over again, so the resonant frequency of the tuning fork similarly changes pitch in a very regular way. I should admit that we've used a trick here—in addition to amplification factors that make Motörhead look like Simon & Garfunkel when it comes to gain levels—to make the signal audible.

The tuning fork we use, in common with many other research groups working on probing single atoms and molecules, is made of quartz and is specially designed to resonate at 32,768 Hz. You have almost certainly encountered one of these tuning forks before: they're the timing element in quartz clocks and watches. Take the tuning fork out of its container, however, and, as Franz Giessibl, a researcher at IBM labs, discovered back in 1999, it's a fantastic sensor for interatomic forces. (Giessibl's introduction of what is now known as the qPlus tuning fork sensor design paved the way for a leap forward in the capability of scanning probe microscopes.) When a probe tip is glued to the end of the tuning fork, the resonant frequency drops (due to the additional mass of the tip) to about 25,000 Hz (25 kHz). But that's not enough to pull the tuning fork output within the audible range. We therefore transposed the signal into the audible frequency range to ensure we could hear it.

We are of course not limited to listening while scanning. We can park the probe over a single atom, as done for the atomic switch experiment, and push it ever closer to the surface. As the force between the tip atom and its neighbor below increases, the frequency drops and the resulting tone sounds not too unlike the "dive-bomb" whammy-bar-on-natural-harmonic effect that metal guitarists love to death. To highlight this, here's a classic metal dive bomb where I've sounded a natural harmonic (at the fifth fret on the G string) and continuously lowered its pitch with the whammy bar:

🔊 https://www.youtube.com/watch?v=_IQcQ4WuQLI.

And here's what the waveform from the tuning fork for the chemical bond formation sounds like:

🔊 https://www.youtube.com/watch?v=3s3sbCxv6r8.

We're listening to the sound of a chemical bond being formed. Well, *okay*, not quite. We're listening to the reaction of the tuning fork as a chemical bond forms: a *single* chemical bond. If I think about this too much I'm always blown away by it. We can hear the formation of a single bond between two atoms by listening to the effect of that interaction on a tuning fork that is literally ten million times larger. The difference in scale is phenomenal. To put it in context, if our friend Marsha was the size of an atom, then the tuning fork would be the size of Earth. And yet that single chemical bond has a measurable effect on the frequency of the fork.

We've activated our switch purely through chemomechanical force. The force of a single chemical bond has been exploited to tilt the switch from the "on" position to the "off" position (or vice versa). And it'll stay that way (at least at the very chilly temperature of 4 K at which we performed the experiment) until we pull it into the other state. Perfectly reliably. The probability of the atomic switch spontaneously flipping to the "off" position from the "on" position, or the other way around, is so incredibly small that we'd have to wait for much, much longer than the age of the universe (around thirteen billion years) to see an uncontrolled flip at the temperature at which the experiment was carried out.

All those atoms are in their places. And we know their positions, right down to a precision of fractions of an atomic diameter. And we also know that they're fixed in place and aren't going anywhere anytime soon.

If everything is predictable to that extent, where does that famed quantum "uncertainty" come in?

To answer that far-reaching question, we'll have to scale back up to the macroscopic world and consider a few classic guitar riffs . . .

Chugging, Fundamentally

It's safe to say that without the guitar "chug," metal music would be a very, very different beast. Virtually every classic (and not so classic) metal song is propelled by the crunchy sound of heavily amplified, palm-muted guitar. From "Holy Diver" to "Holy Wars," from "Am I Evil?" to "The Number of the Beast," the chunky sound of damped strings is at the heart of the rhythmic content, power, and dynamics of metal. What few people realize is that, buried deep in those chugs, the uncertainty principle is hard at work.

Whether or not you're already familiar with the crunchiness of guitar chugging—and, if you're not, don't worry, we'll be looking deep into the anatomy of the chug very soon—let's step back for a second and revisit what happens when a guitar string is not choked and is instead plucked and allowed to ring out, unfettered. I'm going to hit the lowest string on my guitar and let the note sustain for as long as it takes to fade away into the background noise.[10] Here's what the waveform and its Fourier spectrum look like:

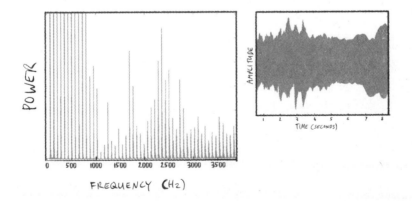

---

[10] In this case it's the opening note of the riff for Soundgarden's "Jesus Christ Pose." (By the way, don't let anyone—least of all the band themselves—tell you that Soundgarden wasn't metal. Go back and listen to that discordant riff, the superlative drums, the rumbling bass, and Chris Cornell's banshee-at-full-moon wailing. Christ, if that's not metal I don't know what is.)

That's a fairly complicated spectrum, but you should expect that by now. The waveform is complex due to the layers of distortion on the guitar signal and the associated rich harmonic structure. Let's cut through that complexity and take a close look at just one of those harmonics. Its frequency spectrum looks like this:

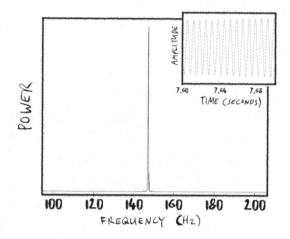

Although we're now looking at just one Fourier component out of that zoo of frequencies, it's nonetheless a cornerstone on which the entire sound is built. The sharpness of the spike in the spectrum means that we have a very good handle on the pitch of this note: it's 147 cycles per second.[11] In other words, we can be rather certain of the frequency of the fundamental.

Now look at what happens when the note is damped (with the edge of the palm of the hand) to give that crunch/chug so beloved of metal guitarists. First, here's the waveform:

---

[11] . . . give or take a few fractions of a cycle per second. The spike is very narrow, but it's not *perfectly* narrow. This nearly-but-not-quite-perfectly-sharp aspect will soon become very important.

TIME DOMAIN

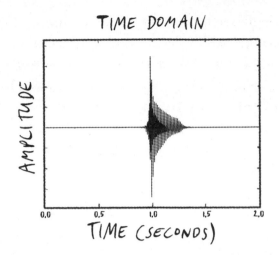

TIME (SECONDS)

As you might expect, it's a much shorter signal. The original undamped note rang out for nearly 20 seconds—until it faded to the audio equivalent of black. This time, however, I've cut the note short by palm muting it. Instead of gradually decaying off over the course of 20 seconds, the "kerrchunk" lasts only a second or so. What effect, if any, will this have on that sharp spike we saw for the frequency spectrum of the fundamental? Let's take a look.

FREQUENCY DOMAIN

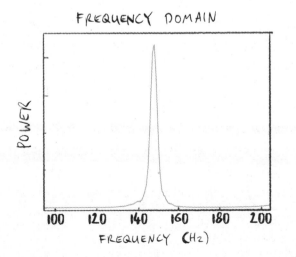

FREQUENCY (Hz)

The peak in the spectrum is centered at the same frequency: 147 Hz. This is not too surprising. It's the same note, after all; it's just been choked so it doesn't ring out. But although the peak is centered at the same frequency, there's an exceptionally important difference when compared to the spectrum for the undamped note. Here's both spectra in the same figure (remember, for simplicity, we're not showing the note's entire frequency spectrum, just that of the fundamental harmonic):

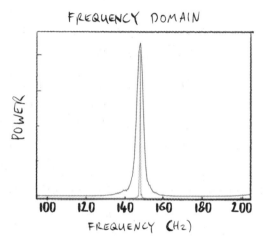

Putting both spectra on the same plot makes it clear that the frequency peak for the undamped note is much sharper than that for its damped counterpart. Both peaks have their centers at the same point on the x-axis, but the spectrum for the damped note is much "fatter."[12]

What we're seeing here is nothing less than the uncertainty principle in action. Given that this effect was the inspiration for both the title and the premise of this book, I hope you'll understand if I spend a little time digging into the minutiae of what's going on here. This won't just be an exercise in the nitpickery and navel-gazing for which physicists are infamous. If you can get your head around why those spectra above

---

[12] Resists temptation to make very weak "if not phatter" pun. (Nu-metal has an awful lot to answer for.)

are a prime example of the uncertainty principle in action, you'll have developed a deep understanding of a key principle in not just quantum mechanics but in *all* of physics.

And you'll have heavy metal to thank for those insights.

The Grand Conjugation

Remember that piece of advice in Chapter 1 for aspiring physics researchers? "If you can't think of anything else to do, try Fourier-transforming everything." You may have noticed that I was at pains when describing the undamped note to point out that it persisted for quite a long time and that this contrasted heavily with the one-second duration of the chug. Yet, in Fourier-space, it's precisely the other way round. The note that persists for a long time has a very narrow frequency spectrum, whereas the short and snappy kerrchunk of the chug is associated with a much wider spectral peak. Put simply:

**Narrow in time, wide in frequency;
wide in time, narrow in frequency.**

This is the uncertainty principle. There's no need to delve into murky quantum mechanical waters—the uncertainty principle is at work all around us on length scales that have nothing to do with quantum physics. It's a very simple concept that is fundamental to the behavior of waves, whether they're large or small, eardrum shattering or inaudible. And central to wave behavior is what's called the conjugate nature of time and frequency.

As we've seen over and over, we can either look at a waveform in time or we can consider its frequency spectrum. The representations are entirely equivalent; they're just different sides of the same wavy coin. When we look at the note that extends for a long period, we find it represented by a sharp frequency spectrum. This means that we're very certain of the frequency of the note. But we're less certain about just *when* that note exists, because it spreads out in time.

Now consider the chug. In this case we have a signal that is much more "confined" in time: instead of the sound droning on for nearly 20 seconds, the note exists for only about a second. We're much more certain of *when* the note sounds—it's only one second, after all. But we can infer from our chug experiment that the frequency spectrum for this much shorter signal will be wider. Or we could simply compare the groundbreaking sustain of the *Tap*'s Nigel Tufnel with the pioneering thrash metal chug of a certain James Hetfield . . .

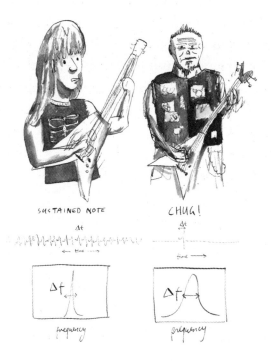

We represent the width of the guitar note, either in time or frequency, using the Greek delta symbol Δ. (As is physics tradition.) The duration of either Tufnel's long, sustained note or Hetfield's signature chug can be written as Δ*t*. In other words, the signal is Δ*t* (pronounced "delta t") seconds long. Similarly, the width of the peak in the frequency spectrum can be written as Δ*f*. The spectrum is Δ*f* ("delta f") hertz wide.

As Δt increases, Δf gets smaller, and vice versa. Remember, from Chapter 3, that frequency is just "reciprocal time." As you can see from the sketch, Tufnel's sustained note lasts a longer time; therefore, the Δt is large, and hence the Δf is correspondingly small. Similarly, as we make a signal more limited in time (à la Mr. Hetfield), its spectrum will become broader.

Just as Tufnel's and Hetfield's styles complement each other,[13] time and frequency are what are called conjugate variables: they each form part of an intrinsically linked pair. The importance of that "conjugation" was realized, once again, by Monsieur Fourier,[14] whose methods allow us to translate a signal from one domain to another. But what does Fourier analysis tell us about *why* the chug has a wider frequency spectrum than its sustained counterpart? To explain, we're going to adopt an approach that is extremely common in physics: we're going to take it to the limits. We'll consider what happens in the most "extreme" situation we can imagine and then wind back from there.[15] You're welcome, metal fans.

---

[13] Metallica paid their dues to the massive influence of "the *Tap*" on their music via the artwork of their fifth, eponymous album.

[14] The precise mathematical definition of what is meant by a conjugate variable is long and involved, but it's not really an oversimplification to state that quantities that can be transformed to each other via Fourier analysis, such as frequency and time, *are* conjugate variables.

[15] First-year physics students are continually asked to show that a particular aspect of quantum physics or relativity theory leads to the classical result "in the appropriate limit." In other words, if quantum theory doesn't return the correct result when we consider the equations in the limit of very large numbers of atoms and/or everyday temperatures—rather than (close to) absolute zero—then there's something wrong. Indeed, this strategy has a special name in quantum theory: the correspondence principle. Similarly, if the equations in relativity didn't produce the results we see in the world around us when we set the speed to 60 miles per hour—rather than 99 percent of the value of the speed of light, say—then the theory is obviously flawed at some level. Think of it as a sanity check.

Notes to Infinity

We will start with the most boring and irksome sound imaginable: a single tone, extending forever both backward and forward in time. Never ending. Never changing. Never relenting. Forget, for now, that the universe/multiverse/alien simulation in which we live had its origins in the Big Bang and may well be heading inexorably toward a Big Crunch, Heat Death, Big Bounce, or some similarly nasty ultimate fate; that there was a beginning and that there'll (possibly) be an end. The tone we have in mind transcends our finitely aged universe—it exists for infinite time. This is a sound that makes One Direction's collected hits almost palatable by comparison, with a sustain that Nigel Tufnel couldn't even begin to imagine. A pure, never-ending sine wave. A perfect—and perfectly maddening—epoch-spanning whistle. Not even "Wind of Change" can compete in the irritation stakes.

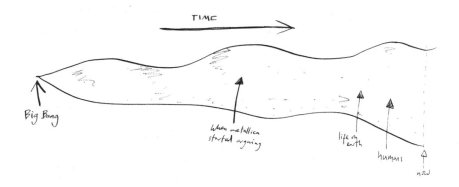

This wave has a width, Δt, that's as big as it's going to get. It stretches backward to minus infinity and forward to plus infinity on the timescale.[16] And as time and frequency are reciprocally related, if Δt extends to infinity, this can only mean that the width of the frequency spectrum is the reciprocal of infinity. And the reciprocal of

---

[16] No, I'm not making this up. This is exactly how physicists think when it comes to Fourier analysis. Infinity is our friend. See the "Maths of Metal" appendix if you'd like to play with this madness a little more.

infinity is an *infinitesimal*: a quantity smaller than the smallest thing ever. A quantity that, by comparison, is dwarfed in size by not only an atom, an electron, or a quark,[17] but is immeasurably smaller than what is widely believed to be the smallest physically possible length scale (on the basis of our current theories): the Planck length.[18] An infinitesimal is the smallest *conceivable* quantity—so small that it could never be measured.

A wave that extends for all time has a frequency spectrum that is the very definition of a perfect pitch. This is the uncertainty principle taken to its very limits: infinitely wide in time, infinitesimally narrow in frequency spectrum. We can't pin down the wave in time because it exists for *all* time. The question "When did the wave sound?" makes no sense. The wave has always, and will always, sound. It's always been there. This represents perfect temporal uncertainty. Yet, simultaneously, we're entirely certain of the frequency because there can be no other frequency: the width of the frequency spectrum is infinitesimally narrow.

Perfect uncertainty in time, perfect certainty in frequency.

To represent that infinitesimally narrow frequency spike, however, we need an entirely new type of mathematical beast, one known to strike fear and foreboding into the hearts of physics students.[19] It's a

---

[17] A quark—as opposed to Quark, the dairy product—is a fundamental particle. Quarks group together, in different flavors, to form hadrons. The most stable types of hadrons are the protons and neutrons found in every nucleus of every atom in (every?) universe. The Large Hadron Collider (LHC) gets its name from the protons that circulate at a speed exceptionally close to the speed of light— just 3 m/s less than the 300,000,000 m/s pace of light. But the LHC doesn't just use protons; it also gets much heavier at times. Heavy metals, such as lead, are also sometimes circulated in the ring to create the oldest form of matter in the universe, something known as a quark-gluon plasma, which is believed to have formed microseconds after the Big Bang.

[18] The Planck length is 1.6×10^{-35} meters. That's roughly 10 million, million, million, million times smaller than an atom. To give you an idea of the difference in scale, we're talking about the width of a guitar plectrum as compared to the distance to the Andromeda Galaxy and back.

[19] You might want to play the opening track of Sabbath's debut album at this point to conjure up the appropriate sense of doom and despair.

monstrosity known as the Dirac delta function,[20] named after a theo-retical physicist who, despite not having the fame and cultural clout of a Schrödinger or a Heisenberg, was at the very epicenter of the quantum revolution. (Not that Dirac would have given a hypothetical cat's ass about fame, fortune, or eponymous memes.)

Delta Blues

His great discoveries were like exquisitely carved marble statues falling out of the sky one after another. He seemed to be able to conjure laws of nature from pure thought.

This was Freeman Dyson's appraisal of Paul Dirac, a physicist many have dubbed Britain's Einstein, who was instrumental in the development of quantum theory.[21] By 1925, at the tender age of twenty-three, Dirac had already laid many of the foundations of the field, and just three years later went on to predict the existence of antimatter. What's astounding is that he arrived at his predictions solely via what many would consider arcane—but which Dirac described as beautiful—mathematics. Eccentric and taciturn, with a fondness for excruciatingly long silences during stilted exchanges with friends and colleagues, Dirac was described by Niels Bohr as "the strangest man I've ever met."[22] If you're a fan of the TV show *The Big Bang Theory*, you might have in mind a certain Sheldon Cooper. But that'd be wide of the mark by a couple of light-years.

---

[20] I *did* give you fair warning all the way back in Chapter 3 that Dirac's delta was going to raise its ugly head, so don't say you weren't given a heads-up.

[21] Dyson himself is no slouch when it comes to contributions to physics. He's a fascinating man whose interests and work span quantum electrodynamics—the theory of how light and matter interact—to the origin of life, space exploration, religion, and beyond.

[22] I thoroughly recommend Graham Farmelo's engaging biography of Dirac, fittingly entitled *The Strangest Man*.

Sheldon certainly shares Dirac's literal-mindedness[23] but is many orders of magnitude more verbose.

In developing his particular breed of quantum mechanics, Dirac needed to exploit a mathematical trick that would allow him to specify a value for a particular quantity at only one point in space (or at one point in time). Dirac introduced this trick in his seminal and influential textbook, *The Principles of Quantum Mechanics*, published in 1930, but Fourier had been aware of a similar scheme over a century earlier (as had a number of his contemporaries, including his compatriot and fellow mathematician, Augustin-Louis Cauchy). Nonetheless, it's the twentieth-century British theorist who garners the credit for the introduction of what is now universally known as the Dirac delta function.[24]

Dirac's delta is, to put it mildly, a bizarre concept. It's a mathematical "function" that is zero everywhere except at a single value. Let's return to the frequency spectrum for that epoch-spanning pure tone we were discussing.

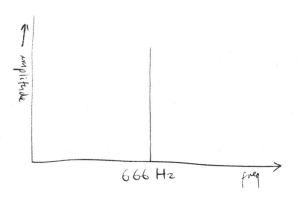

---

[23] Farmelo recounts that Dirac once asked Heisenberg just why he liked to dance. He was told that it was a pleasure to dance with nice girls. After roughly five minutes of ruminating about this, Dirac turned once more to Heisenberg. "But how do you know beforehand that the girls are nice?"

[24] "Function" is not really the correct way to describe Dirac's delta. It's not a function in the widely understood meaning of the term. For the mathematicians and physicists who might be reading, the Dirac delta only makes sense when it's saddled to an integral.

The frequency spike sketched on the previous page is represented mathematically by Dirac's delta. The wave has only one, very specific, value. That alone might not sound too weird, but the devil, as ever, is in the detail. Just what do we mean by "one, very specific, value" of frequency?

Is it 66.60 Hz?

Or is it the more precise 66.6000 Hz?

Or is it the even more precise 66.6000000000000000000000 Hz?

Or is it the exceptionally precise 66.600000000000000000000000 000000000000000000000 Hz?[25]

How do we define that single specific value of frequency? What precision do we use? How many decimal places do we need to *precisely* define that value of frequency, beyond a shadow of a doubt? How do we define that value with total, utter certainty? There's only one way of doing that mathematically when it comes to our frequency spectrum and that's through Dirac's delta function.

Let's consider this concept of a perfectly precise number in a slightly different context. The motley crew on the next page are all metalhead "mathemusicians" auditioning for a Slayer tribute band, Angle of Death. All the other roles in the band have been filled—it's just the vocalist/bassist, Tom Arayabhata,[26] who remains to be selected. The other members of the band are sticklers for precision, so not only do they want Angle of Death to sound like the real deal, they want to look like Slayer, too. Alongside assessing a variety of other physical characteristics, they

---

[25] In physics and engineering, the concept of *significant digits* is extremely important. Zeroes trailing a number to the right of a decimal point are assumed to be significant. They tell us about the precision of the value (because why would we bother to write them down if they didn't signify something important about the number?).

[26] Aryabhata (~475–550 CE) was an Indian mathematician and astronomer who had a major influence on many aspects of physics and mathematics. He's a particularly important figure in the context of this book, given that he not only defined sine, cosine, and other trigonometric functions back in his day but collated tables of their values (to four decimal places) for angles between 0 and 90 degrees.

measured the height of each person who auditioned and, being mathematicians, they plotted a histogram (otherwise known as a bar chart) to help them make their decision.

The histogram shows the distribution of heights among the Araya wannabes. But how many of those auditioning are exactly, precisely six feet tall?

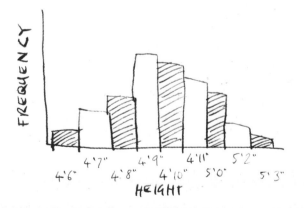

Answer: zero.

To be exactly, precisely six feet tall in a mathematical sense means that your height must be expressed with infinite precision. No one auditioning for that Angle of Death gig is precisely six feet tall. Indeed, no one *in the world* is precisely six feet tall. Although we'll sometimes say that a particular person is "exactly" six feet tall, we don't mean that they're six feet tall with a precision that runs out to an infinite number of decimal places: "6.0000000000 . . ." Practically, we can only consider someone's height falling within a particular range of values (or being larger or smaller than a particular value).

The Dirac delta function, however, allows us to select out that single, infinitely precise value. It amounts to taking one of the bars of the height histogram and shrinking its width until it is infinitesimally narrow: as thin as thin could ever be.[27] In the perfect mathematical world of Dirac's delta, someone *can* be precisely six feet tall. And a headbanger's flowing locks can be precisely three feet long. And a wave can have a frequency that is infinitely precise.

But we of course don't live in a world of mathematical perfection.[28] It's about time we got back to the imperfections and uncertainties of our real, physical universe. We metal-loving experimentalists much prefer our science—and our music—warts and all.

Masters of Reality

Before moving on, and at the risk of belaboring the point to death—or, indeed, to deth—let's take a second to recap. First, in the perfect world of pure mathematics, we can have a wave that lasts for an infinite amount of time. This means that we have absolutely no idea when the wave started or when it will stop (because it didn't and it won't—it's always been there and always will be there). In other words, we're per-

---

[27] At the risk of causing a migraine—I did say that the Dirac delta function was something of a mathematical monstrosity—the "bar" simultaneously gets taller and taller as it gets thinner and thinner. When it's infinitesimally wide, it's also infinitely high. This means that the area of the delta spike is 1. Although this is an exceptionally important aspect of the delta function for the detailed mathematics of Fourier analysis, we can fortunately relegate it to a footnote here.

[28] Although it's always worth bearing in mind another well-worn quote from Monsieur Fourier: "Mathematical analysis is as extensive as Nature herself."

fectly uncertain when it comes to the "when" of the wave. When it comes to its frequency, however, we've got perfect certainty because the wave is the purest of tones there could ever be. Dirac's delta function allows us to mathematically represent that perfectly defined frequency as an infinitesimally narrow spike: the ideal spectrum.

That's how the uncertainty principle works for a purely mathematical, infinite wave. But what about the pressing issue of those chugging guitars that have been reverberating for the last few pages? Well, we've been to the limit. Now we're going to slowly walk back from the edge of infinity.

Instead of letting our perfect wave run for all time, let's make it more realistic and confine it to a well-defined period. You know that when we let the guitar note decay naturally (rather than using palm muting to produce a chug), it took about 20 seconds to fade out. We're similarly going to very lightly damp our hitherto infinite wave so that it decays slowly and is gone by $t = 20$ seconds, just like the guitar note naturally fading out after the string has been plucked. Remember that outside the idealized, rarefied universe of pure mathematics, back in the real world of friction, finite time, and Fall Out Boy, the string will lose energy and eventually come to rest, even without putting the palm of our hand on the string to mute it—for one thing, it's knocking air molecules back and forth as it moves. As discussed back in Chapter 2, this natural damping will cause the oscillation of the string to die away.[29]

We mathematically model this damping effect by multiplying our infinitely prolonged perfect sine wave by what's called an exponential

---

[29] There *is* a way to coax an if not quite infinite, then dramatically prolonged, sustain out of a guitar, however. It's a neat little device called an E-bow and it involves pumping in energy at just the right frequency to sustain a note indefinitely (or until the batteries run out or a cramp sets in). The E-bow is not used too widely in metal, but Opeth has put the device to exceptionally good use on a number of their albums. "Bleak," from their genre-defining *Blackwater Park*, released in 2001, exploits the wonderfully evocative sound of the E-bow throughout. My favorite E-bow moments, however, are from Steve Rothery on Marillion's *Misplaced Childhood*—a master class in applying the technique.

decay, as in the sketch below. The wave slowly "fades to black" in just the way we want.[30] (I'll note in passing that the technical term for our fading-out guitar note is a damped harmonic oscillator. That exponentially decaying sine wave is representative of behavior that's the bedrock of *so* much science and engineering, spanning architecture to cosmology and practically every field in between. But let's keep focused on the core issue here: the fundamental significance of the quantum-metal interface.)

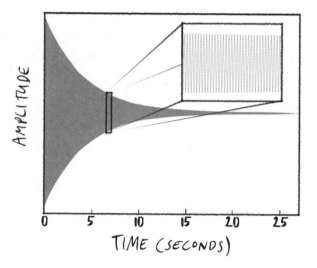

So that's our wave in the time domain. But what about that other, equally valid, conjugate representation? What does the frequency spectrum look like? We know that if we have the sine wave by itself—a note stretching to infinity—it's just a single frequency: a spectral spike. What happens to the frequency spectrum of that perfect wave when it is allowed to decay to nothingness?

---

[30] Note to physicists: Yes, I know I'm simplifying here and that adding damping will affect the resonant frequency of the motion. Again, please let it go! That's a very small change in the context of the lightly damped situations we're considering here.

Here it is:

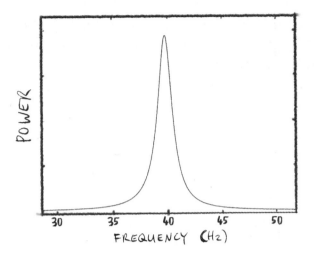

We've narrowed the wave in time. And though we are working with a perfect, idealized, distortion-free note, look what's happened to the frequency spectrum—it's broadened. Narrow in time, broad in frequency. There's the uncertainty principle[31] stripped down to its core. If you come away with nothing from this book other than this concept, I'll consider the many hours I spent writing a roaring success.

Our Δt has dropped from a beyond-epoch-spanning infinite duration to the very much more finite value of about 20 seconds. Let's continue to curtail the note. Instead of 20 seconds, let's have it fade out after only 5 seconds, then 1 second, and then finally we'll "choke" it so that it's only a tenth of a second in duration. You can probably guess at what's going to happen to the evolution of the frequency spectrum.

---

[31] Note that there's been nary a mention of an observer throughout the discussion of the uncertainty principle. Nor have zombie cats clawed their way into the explanation at any point. I'll return to this lack of traditional quantum memes at the end of the chapter.

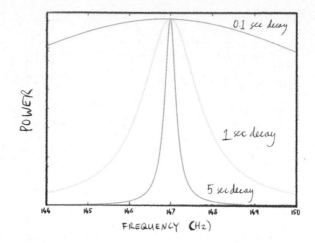

Again, as we make the note progressively shorter in time, its frequency spectrum gets wider.[32] This can only mean one thing. More Fourier components—that is, more frequencies—are required to represent the shorter kerrchunk sound. But . . . that's a little perplexing. It's the same fundamental tone, after all. *It's the same note.* All we've done is make that note shorter by damping it more heavily—moving from a "kerrang" to a kerrchunk. Why should simply making a note shorter mean that more frequencies are required in the spectral mix?

Summing Up

Let's step back and recall the basic principle underpinning Fourier's analysis technique: we can take any waveform, signal, or function[33]

---

[32] I've got to come clean and admit that I've invoked a little bit of graphical "trickery" here. I've scaled all of the frequency spectra to the same height. In addition to each spectrum widening as the length of the wave gets shorter, the height of the peak also gets smaller (in order to keep the area under the peak the same). I've not shown this decrease in peak height, however, because if I simply plotted the graphs without any adjustment, it's more difficult to see the change in width.

[33] Just in case the mathematicians start getting restless, I mean *practically* any waveform, signal, or function.

and break it down into a sum of sine waves with different frequencies, amplitudes, and phases. From this perspective, it makes perfect sense that the spectrum for the decaying guitar note contains many more frequency components than the imagined never-ending one. If you look back at page 176, at the waveform of our perfect note as it decayed over 20 seconds, you'll see at a glance that the height and the shape of the wave changes; the amplitude decreases as the note fades away. The fundamental concept at the core of Fourier's method dictates that the spectrum *must* contain additional frequencies for any signal that isn't a pure, solitary sine wave. The spectrum tells us which particular frequencies we need to add to our mix in order to construct the decaying signal. It's a "deconstruction" of the signal into its Fourier components. Only sine waves with the correct frequencies and amplitudes,[34] when added together in the right way, will reproduce the decaying signal. And Fourier analysis tells us how to work out just what those frequencies and amplitudes need to be.[35]

Fortunately, however, we don't need to know how to do the math to grasp the concept. It's just an extension of what we already know. We can generate a set of sine waves—sounds, in this case—with the correct frequencies and amplitudes and feed each signal to its own dedicated amplifier.[36]

It'd ideally look a lot like this:

---

[34] And phases. Although we don't need to worry about phase just yet. See the following chapter.

[35] For those interested in all the gory mathematical details, see the appendix.

[36] It's as easy as pie (or even pi) to generate sine waves these days. Grab your laptop, iPhone, iPad, or whatever piece of sound-generating technology is on hand—there are many apps around that will generate a pure sine tone, including Audacity. (By the way, if *Marshall* is listening, I would just *love* to do this experiment. I don't have quite enough amplifiers at hand at the moment . . .)

That wall of sound won't quite recover the exact signal. But it'll be a damn good facsimile. And the more sine waves we use, the more faithful the reproduction will get. If we pile Marshall upon Marshall, each generating its own sine wave, we'll get closer and closer to a replica of the sound we want. *Any* sound we want. In principle.

What we're doing with that fictional bank of Marshalls is something called additive synthesis. It's nothing more than the application of Fourier's mathematics to the generation of music. Now used widely in synthesizers, the idea was around a long, long time before synths sprang to life—or were spawned from the foulest depths of Hades, depending on your perspective—in the 1970s. Many centuries before Fourier worked out his mathematical analysis techniques and appreciated their ramifications, additive synthesis was being unconsciously exploited by pipe organists.

In the Middle Ages—and, of course, to this very day—huge pipe organs were equipped with "stops" that could be pulled out to combine the sounds from different pipes. Unbeknownst to the organ player, they were performing sophisticated Fourier synthesis of waveforms long before Fourier was born. By manipulating the stops, different harmonics were added together to change the tone and timbre of the resultant note. The pipe organ is the medieval equivalent of our Marshall überstack.

ART AND APPARATUS

A spectacular marriage of heavy engineering and music technology in the late nineteenth century, known as the Telharmonium, can lay claim to being the world's first synthetic instrument. It generated sounds *electromechanically*. Musical tones were produced from spinning tone wheels. These are very similar to gears in that they are rotating disks with cogs. The cogs occur at a regular spacing along the circumference of the wheel, which is spun in front of a pickup rather similar to those found in electric guitars. The entire setup looks like this:

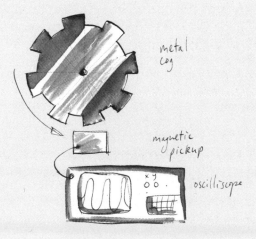

As the regularly spaced cogs pass in front of the pickup, they generate an alternating current—a sine wave variation in current—in the coil, which in turn produces a particular tone when fed to a loudspeaker. (Except that loudspeakers as we know them didn't exist at the time the Telharmonium was invented.) Rotate the disk at a higher rate and the pitch of the note increases. It's a very elegant and clever idea. With a set of tone wheels all rotating at different speeds, different tones can be generated and mixed. The Telharmonium was, in essence, the first electrical synthesizer. In common with that fictitious bank of Marshalls, the Telharmonium synthesized sounds by adding together different Fourier components: individual sine wave signals created by tone wheels spinning at different speeds were summed to produce an array of sounds. We'd call this Fourier synthesis—because that's exactly what it is—but, as mentioned above, a more popular term in sound engineering/music technology for the process powering the Telharmonium is additive synthesis.

Invented by the melodiously monikered Thaddeus Cahill in the very late nineteenth century, the first version of the Telharmonium weighed 7 tons. Telharmonium Ver 2.0, however, weighed in at *200 tons*! The control console looked less like a keyboard and rather more like NASA's Mission Control. The Telharmonium itself was so large it was housed in the basement of the concert hall, with the musician "driving" the console upstairs in front of the audience. Lengthy cables were fed through holes in the floor of the theater, connecting the console with the Telharmonium "mother ship."

The tone wheel technology of the Telharmonium evolved and decades later became responsible for the signature sound of the Hammond organ. This in turn meant that Fourier's "sums of sines" approach to sound generation was the foundation of the music of a band that was in turn responsible for helping lay the foundations of heavy metal: Deep Purple. For once, it's not the guitarist who takes center stage in the story. (Despite Ritchie Blackmore's influence on generations of metal musicians. Who doesn't learn the monolithic riff to "Smoke on the Water" very early in their guitar-playing career?) Powering the Purple sound were the virtuoso contributions of keyboardist Jon Lord, arguably the most famous proponent of the Hammond organ in rock (and certainly *In Rock*).

Lord fed the output of his Hammond organ to Marshall guitar amplifiers to produce the growling signature tone that battled with Blackmore's guitar on "Space Truckin'," "Highway Star," "Perfect Strangers," and so many of the other classics in the Purple canon. Instead of having that

wall of Marshalls we imagined a few pages back, each producing its own sine wave, the Hammond carries out Fourier synthesis by mixing the various tone wheel signals
together to produce the final gut-rumbling sound. Adding
the Marshall distortion on top, à la Lord, further increases
the harmonic complexity.

In these digitally enabled times, of course, we don't need to use a
Victorian era Telharmonium, or a Hammond organ, or a wall of Marshalls (or any combination thereof) to sum up signals. We can easily do
Fourier synthesis on a computer. This is not as much fun as dealing with
200 tons of equipment, but it's precisely the same underlying concept:
we build up a signal from a set of different sine waves.[37] And the uncertainty principle tells us that if the signal doesn't last for long, we're going
to need a wide range of sine waves to represent it.

As we move from a sustained note to a chug and the frequency
spectrum broadens, we'll need to use a larger number of signal genera-
tors—whether tone wheels in the nineteenth century, or simply more
digital signals on a laptop/iPad in the twenty-first—covering a broader
range of frequencies, to synthesize our kerrchunk.

How far can we take this? What if we have the shortest possible note
ever? A note so short that it's beyond the capabilities of even the fastest
shredder to play. A chug that lasts for an infinitesimal amount of time.
What then does the frequency spectrum become?

We've seen that an infinitely sustained note—that never-ending
whistle—has a perfectly narrow frequency spectrum: a delta function

---

[37] I need to take a moment to point out a subtle, but fascinating, difference
between the digital domain and the universe of pure mathematics. (Our digital
reality of course didn't exist when Fourier first put nib to paper. Fortunately.
There's no telling how his work might have suffered if he'd been checking his
Facebook and Twitter feeds every ten minutes.) In the digital world we can't
have infinitesimal quantities—a computer can only store numbers to a finite
precision. Thus, every digital representation of a mathematical function, like a
sine curve, is an approximation.

spike. Remember that all-important reciprocal relationship between frequency and time that's the core message of the uncertainty principle: infinitely wide in time means infinitesimally narrow in frequency. *And vice versa.*

So, if we have an infinitesimally short chug—in other words, a delta function spike in time—its frequency spectrum will be *infinitely* wide. To create the shortest possible sound, we need the widest possible range of frequencies. And we need to ensure that the infinity of sine waves all line up in precisely the right way so that they cancel out everywhere but for that infinitesimally short moment in time. If we put this in the context of the uncertainty principle once again, what Fourier tells us is that complete certainty as to when a note sounds translates to complete uncertainty about the frequency of that note. The uncertainty is so large that an infinitely wide band of frequencies is required to represent the note.

Walking back from infinity once again, but this time in the frequency domain, as we narrow the width of the spectrum we'll find that our signal in time will broaden out. The longer a note sustains, the narrower the frequency spectrum becomes—as in the Tufnel sustain vs. the Hetfield chug.

Watching the Watchers?

As we close this chapter on the origin and ubiquity of the uncertainty principle (on all scales), I'd like to note that not once in the preceding pages has the quantum "observer effect" raised its ugly head. The observer effect, simply put, is the idea that measurement disturbs a system at the quantum level. This (very real) effect has too often been conflated with Heisenberg's principle, leading to the difficult-to-shake, incorrect notion that uncertainty arises from measurement: we measure, we disturb, and therefore we can never be entirely certain of the state of anything because of our influence. But as we've seen repeatedly in this chapter, the uncertainty principle arises naturally when we consider the mathematics and physics of waves, regardless of the scale on which they appear. It's always there, regardless of how accurately, or how delicately, we can make a measurement.

The uncertainty principle, therefore, isn't really a quantum effect per se; there's nothing weird or esoteric about it. It's a signature characteristic of waves—an innate property that is no more weird than the relationship between frequency and wavelength. In this sense, it's as obvious as $2 + 2 = 4$. Or $600 + 60 + 6 = 666$.

The observer effect is regularly confused with the uncertainty principle in not only popsci books and articles, but in a variety of physics textbooks. This is rather misleading, but hardly surprising, because even Heisenberg himself mixed up the observer effect with the physics and maths underlying his principle. The uncertainty principle is just a natural consequence of imbuing matter with wavelike characteristics. As we've seen, once we have waves—be they at the level of single atoms and molecules, or at the much heavier metal end of the scale—the uncertainty principle reigns.

Just what is meant by an "observer" in quantum theory is the subject of a great deal of mythology and mysticism. Too often, the "observer" is confused with consciousness or subjectivity. This is where quantum woo comes up with much of its "reality is all an illusion we create" claptrap. Yes, you change things by "observing" them . . . alas, so does any object, even an entirely inanimate one. A rock scatters quantum waves; it's an observer. A Marshall cabinet scatters quantum waves; it's an observer. A guitar plectrum—at rest or in motion—scatters quantum waves; it's an observer. And a box in which a cat is trapped in a famous *gedankenexperiment* is similarly an observer. Long before that box is opened, Schrödinger's infamous imaginary feline has been observed gazillions of times.

Much of the weirdness of the quantum world arises only in experiments that are carefully engineered so that matter waves remain in step and interfere constructively and destructively in just the right way.

Even the most serene and quiet of environments, let alone a metal gig at 120 dB and beyond, is enough to disrupt those quantum waves. This disruption counts, in essence, as a measurement. Consciousness is not required. Our environment is full of rock-stupid objects, big and small, that can scatter the waves and confound—or at least complicate—our attempts to analyze them.

This wave scattering is known in the trade as *decoherence*. Waves are *coherent* when they're in step with each other: the peaks and troughs line up. And as we touched on back in Chapter 3, we use phase to measure the extent to which waves stay in sync. Given just how important phase is to the quantum world, therefore, it's about time we bite the bullet and look at the concept in some detail. The best way to do this is to enter another dimension.

Chapter 8

INTO ANOTHER DIMENSION

The earth is a very small stage in a vast cosmic arena.

—Carl Sagan[1]

---

[1] *Pale Blue Dot: A Vision of the Human Future in Space*, Random House, 1994.

A major part of the appeal of metal is the overblown, dramatic, and histrionic quality of not just the music but the entire genre and subculture. The onstage performances and persona of a Dickinson, Dio, Halford, or Hetfield aren't exactly those of a shoe-gazing shrinking violet who shuns the limelight. As Bruce Dickinson, whose energetic and engaging fronting of Iron Maiden plays a key role in their massive popularity, put it:

> Standing there looking beautiful is not performing. Performing comes from engaging. And it's hard work; you've got to work to do that. Audiences are brutal—I mean, they will eat you alive—because they all paid good money to be entertained, and they turn up expecting to be told what to do. They may disagree with that, but actually they do. They turn up to a show, and they go, "I came here to be . . ." you know. "This is gonna be good. What do I do now?"
>
> And the band guide[s] them; you guide them to a great place, where they go, "Wow! What a great experience. Wasn't that cool? And we did this. And we all sang, and we all chanted." But if the guy comes up on stage and goes, "I'm gonna stare at my shoes and it's all about me," they're gonna kill you. 'Cause it's not all about you, dude—it's about the music, and it's about them; it's about the audience.[2]

Coupled with the drama of the live performance is the striking iconography of metal. As with so many other aspects of the genre, Iron Maiden once again leads the way. Eddie the Head—or, more accurately,

---

[2] From Katherine Turman's October 2015 interview of Dickinson. UltimateGuitar.com: www.ultimate-guitar.com/news/general_music_news/ bruce_dickinson_standing_there_looking_ beautiful_ is_not_performing_ performing_comes_from_engaging.html.

"Eddie the 'Ead"[3]—is *the* iconic image in metal, an omnipresent character throughout Maiden's career, featured on every album cover and in every live performance. Eddie has built the Maiden brand, a brand so strong that even Lady Gaga, style icon supreme, has been photographed while bedecked in an Iron Maiden T-shirt.[4] The Eddie figure, created by Derek Riggs, is instantly recognizable, despite its evolution over the years from terrifying punk on Maiden's eponymous debut back in 1980 to the Mayan theme of their 2016 release, *The Book of Souls*.

Although not every metal band is fortunate enough to have an Eddie—and the associated almost limitless marketing and merchandising opportunities—there's nonetheless a huge emphasis on defining a brand and identity via a striking logo. The spikier the logo, the better. Death metal bands, in particular, have a penchant for exceptionally spiky logos that are virtually indecipherable. Attempting to decode the logo to find the band name is often an exercise in headache-inducing, make-my-eyes-bleed masochism. (This is death metal, after all.) My prize for most-difficult-to-decipher logo goes to Cattle Decapitation (a deathgrind quartet from San Diego that, following in the footsteps of a band like Carcass, writes about the mistreatment and consumption

---

[3] The tireless driving force behind Iron Maiden, Steve Harris, is a native of the East End of London. As with accents in many other areas of England, a signature characteristic of the East End vernacular is the deletion of the voiceless glottal fricative. Or, if you'd prefer that in plain English: Eastenders have a habit of dropping their *h*'s. The pronunciation of the letter *h* has also been described as a shibboleth (at least in the Wikipedia entry for the letter) in the context of Northern Ireland, with Catholics using "haitch" and Protestants using "aitch." I grew up in a border county in Ireland (Monaghan) and must admit that as a child and teenager I was never aware of this. But then, I'm not the most observant. The first time I really became aware of the vagaries of my accent was when I received the set of student feedback for the first undergraduate lecture course I gave at the University of Nottingham. There were a number of references to "dirty tree and a turd." (To close the loop on this footnote, Harris has regularly been spotted wearing a T-shirt that reads "Whale oil beef hooked." Try saying that quickly . . .)

[4] *The Number of the Beast*, no less.

of animals[5]), but there are hundreds, if not thousands, of exceptionally pointy and virtually unreadable motifs out there. This poster for Bay Area Death Fest 2 (2015) says it all.

Among all that spiky metal madness, did you spot *that* logo? It takes a special combination of marketing genius and cheeky sense of humor to subvert death metal conventions to the extent managed by Party Cannon. Their bubbly logo stands out like a sore thumb—or, given that this is death metal, like a mangled, rotting, and pus-covered appendage—in the roll call of spiky, stark, bleeding-eye motifs on that poster. It's a great shame that this book can't include the color version of

---

[5] Both Bill Steer and Jeff Walker of Carcass are vegetarians. (Steer was a vegan for many years.) Other metallers of note who are vegetarian or vegan include Mark "Barney" Greenway of Napalm Death, Bill Ward and Geezer Butler of Black Sabbath, Chris Adler of Lamb of God, Devin Townsend, and Rob Zombie.

the poster, with their logo in all of its cheerful glory[6]—that really would have hammered home the contrast. Party Cannon hails from Dunfermline in Scotland (which might go some way to explaining their rather refreshing take on death metal iconography[7]), describe themselves as "cute boys who like weird music," and call their genre-defying output "death metal/party slam." When the band reaches its (un)natural conclusion, they'll have a glittering career in innovative advertising and marketing ahead of them.

Party Cannon's cartoonish subversion of death metal clichés is refreshing, but hardly high art. Yet a significant number of bands have embraced an ethos where their music and iconography are all about pushing artistic boundaries, to the point where the criticism perhaps most dreaded by a metal musician—that of pretentiousness—has occasionally been leveled. (To many fans, metal is all about "keeping it real"; an accusation of pretentiousness is therefore a death knell—and not in a good, metal way—in certain circles.) Although it's usually the prog metal stylings of Tool that are dissected in the context of metal "artiness," the band is notoriously reclusive and not often inclined to open up about its creative process. Long before Tool appeared on the scene, however, the "metal as art" theme was a key motivation in the development of the music, image, and presence of the innovative Swiss band Celtic Frost, to whom the term *avant-garde metal* has often been applied.

H. R. Giger was a Swiss artist perhaps famed most for his dark and disturbing design work on the *Alien* movies. Giger's signature and surreal biomechanical style struck a chord with, and was adopted by, many metal (and rock) artists including Emerson, Lake & Palmer—ELP—(whose 1973 *Brain Salad Surgery* album famously featured Giger artwork), Danzig, the aforementioned Carcass, Dead Kennedys, and

---

[6] Like this: twitter.com/AstroKatie/status/736375968235802624. (By the way, if you're on Twitter you owe it to yourself to follow Katie Mack.)

[7] The Scottish have been stereotyped for too long as dour and humorless. One name is enough to put the lie to that particular myth: Billy Connolly. A consummate stand-up comedian, Connolly's Glaswegian wit, barbed humor, and perfect comic timing have had audiences in tears of laughter for decades.

Korn. But the strongest connection between the metal genre and Giger himself was via Celtic Frost, who actually collaborated with him. Giger's *Satan I* painting was used for the cover of Frost's second album, *To Mega Therion*, released in 1985, but the links between Giger and the band go much deeper than this, as described by Tom Gabriel Fischer, the creative force behind Celtic Frost, in a statement to the media shortly after Giger's passing in 2014:

> *H. R. Giger became our mentor, against all odds, when we, somewhat audaciously, first established contact with him some thirty years ago. At a time when almost everybody ridiculed, ignored, or even obstructed the music the then almost completely unknown Swiss underground band Hellhammer was creating, Giger listened to us, talked to us, and gave us a chance. Not least at a time when he was at one of many peaks of his path.*[8]

The idea of metal as highbrow culture might seem ludicrous to many outside the subculture. Then again, metal and quantum physics may not have seemed like a natural pairing when you picked up this book, but I'm hoping your perspective on that has changed somewhat by now. Maybe metal as *avant garde* art is similarly easier to swallow at this point.

A Dimensional Shift

All this talk about the art of metal is going somewhere, believe it or not—we're about to make a leap from audio to the visual. And in doing so, we're taking Fourier's analysis up a notch. Thus far, we've seen how Fourier synthesis and "desynthesis" can be used to convert between time and frequency. Or, in other words, between time and reciprocal time. Now we're going to consider patterns in space and *reciprocal space*. Instead of focusing on the analysis of the frequency components that

---

[8] http://teamrock.com/news/2014-05-13/former-celtic-frost-men-pay-tribute-to-h-r-giger.

build a riff or a guitar solo, we're going to deconstruct an *image* into its Fourier components. From gory metal album covers to Hubble Space Telescope images, the type of mathematical analysis we're about to dissect is truly universal in its scope.

Like all aspects of Fourier's approach to analyzing patterns in the mathematical and physical universe—or multiverse, if you're so inclined—reciprocal space, despite the perhaps daunting term, is an elegant, powerful, and, as we're about to see, simple concept. To bring out this simplicity and elegance we're going to enlist the services of a band that, much like the Scottish deathcore act Party Cannon, firmly went against the grain of metal conventions: Stryper.

Stryper formed in 1983 and, at least at the time of this writing, are still recording. They're unusual, if not quite unique, in the metal scene because they're a Christian band; in sharp contrast to the traditional metal stylings of the number of the beast, pentagrams, and the ubiquitous "horns," Stryper reinvented and reinverted the symbolism, embedding their Christianity in all areas of their act. The symbolic "666" became "777" (the number seven has a special place in Christian mythology—seven archangels, seven deadly sins, seven virtues, etc.), Bibles were thrown into the audience, and the crowd was encouraged to hold a "one way" finger, rather than the horns, aloft during Stryper's gigs. But it's their onstage costumes that I want to focus on here. Stryper is going to help us see the light when it comes to reciprocal space.[9]

The Stryper moniker and logo come from the following biblical verse (to which they made regular reference): "But he was wounded for our transgressions, he was bruised for our iniquities: the chastisement of our peace was upon him; and with his stripes we are healed" (Isaiah 53:5). They took the "stripe" idea and, ahem, ran with it . . .

---

[9] And if you're still puzzling about that joke that opens Chapter 7, don't worry. The Stryper guys' choice of stage wear is going to help out there, too. Position, momentum, reciprocal space, and the uncertainty principle are all intimately related and, as we'll see, can neatly be explained via Stryper's sartorial selections.

You are really missing the full effect, believe it or not—the stripes in question are a searing yellow and black.[10] To hammer home the concept, Stryper's debut album, released in 1984, was entitled *The Yellow and Black Attack*. It certainly was.

Remarkably, Stryper's dizzying stripes are really all that we need to explain the concept of reciprocal space. Let's consider the spandex strides of drummer Robert Sweet in the cartoon below:

---

[10] You can see them in all their godly glory here: http://mikesdailyjukebox.com/strypers-first-new-video-in-20-years/.

Let's say, for sake of argument (and to keep the numbers simple), that Mr. Sweet's inside leg measurement is 70 cm. The stripes along his leg form a repeating pattern over this length; the stripe-to-stripe separation defines a spatial period. This is nothing more than the spatial equivalent of the period of a sound wave we've seen time and time again in previous chapters. In this case, instead of the period being measured in seconds (or milliseconds, or nanoseconds), it's given in centimeters (or meters, or millimeters, or nanometers . . .). We're working with spatial, rather than temporal, units.

The cartoon on the previous page is, however, a two-dimensional representation of Sweet's stripes. When considering sound waves, we've been analyzing one-dimensional functions: the volume (or amplitude) varied just as a function of time. Mr. Sweet's trousers are a different matter: the cartoon spandex has both length and breadth. We can, however, strip the problem down to one dimension (while preserving Mr. Sweet's modesty) to arrive at a better comparison with the sound waves we've been considering.

Let's draw a vertical line along one of the legs of the spandex trousers, as shown in the top image below. If we plot the variation in the brightness of the stripes along this line, we get the function shown on the right in the bottom image. This is a square wave, rather than a sine wave, but that doesn't matter; it's still a simple repeating pattern.[11] And because it's a repeating, periodic pattern, we can define its period simply by working out the repeat length.

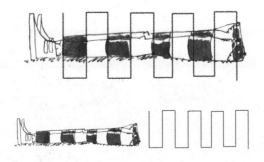

---

[11] We're going to see how to synthesize the Stryper stripes from sine waves very soon, in any case.

Let's call this period T_s (where the subscript s is short for "space," in order to highlight that this is a spatial, rather than a temporal, period).

And now we're going to make the conceptual leap to reciprocal space. (But I've got to stress that it's not really a leap; it's at most a gentle hop, skip, or jump.) The spatial period of the stripes is simply the distance from one stripe to another: 10 cm. If we can calculate a spatial period for the stripes, then we can also calculate their spatial *frequency*: a number describing how often the pattern repeats. We know how to do this from previous chapters. It's easy. If the period, T, of a sound wave is 0.01 seconds, then its frequency, f, is $1/T = 1/0.01 = 100$ Hz. The wave repeats one hundred times in a second.

We do exactly the same thing when it comes to spatial frequency. The spandex stripe pattern has a period, T_s, of 10 cm = 0.1 m. That means its spatial frequency, $1/T_s = f_s = 10$ per meter. The physics shorthand for "per meter" is m^{-1}, i.e., 1/meter. (We're accustomed to using Hz for the frequency of a sound wave, but this is nothing more than a "per second" unit. 1 Hz = 1 per second. 100 Hz = 100 per second. Or, to use that physics shorthand again, 1 Hz = 1 s^{-1}; in words, "one per second.")

The term "reciprocal space" arises from nothing more than this basic relationship between period and frequency. Frequency is reciprocal time and is measured in units of 1/seconds; *spatial* frequency is a measure in reciprocal space and has units of 1/meters. The stripe pattern on Sweet's spandex has a spatial frequency of 10 m^{-1}. If we were to make the separation of the stripes smaller, then we could pack more stripiness into a length of one meter, so the spatial frequency would increase; increase the spacing between the stripes, and the spatial frequency goes down.

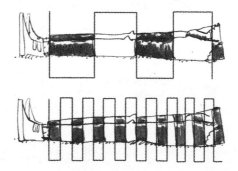

That's the core concept at the heart of reciprocal space: we think in terms of spatial frequencies rather than spacing and distances. Just as we've seen before for the period and frequency of all those metallized sound waves we've explored, we can choose whether to consider Stryper's stripes in terms of their spacing or their spatial frequency; they're entirely equivalent representations of the pattern.

The stripes are a pattern in space rather than time. But they're a pattern nonetheless. And that means, as for any other pattern in time or space, we can apply Fourier analysis. Just as we saw (and heard) notes on a guitar string, we can break the pattern down into a sum of sines: a sum of harmonics. In the inset on the left-hand side of the figure below, I've plotted the first harmonic in both one dimension and two dimensions. (The 2-D pattern is simply lots of 1-D patterns lined up alongside each other.[12])

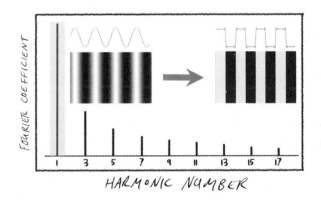

HARMONIC NUMBER

The first harmonic by itself is just a pure sine wave. It captures the correct spatial frequency—the peaks and troughs line up with the lightest and darkest points, respectively—but it's clearly not the right pattern. A sine wave is rather different from a square wave: a smoothly

---

[12] There's an important, but subtle, point here. What we're actually looking at in the illustration is just a *section* of the sine wave. To capture the wave for a single delta-function frequency spike, this book would have to be infinitely long. (And I am sure that there are those who think it is quite long enough as it is, thank you very much.) Remember from Chapter 7 that an infinitesimally narrow spike in frequency means that the corresponding sine wave endures indefinitely.

varying curve vs. an abrupt "on-off" signal. This is the visual analog of the *Sanitarium* analysis we covered in the first couple of chapters, specifically the difference between our E note when whistled and when played on a guitar. The representation above is the visual equivalent of our pure whistled E. It's the right fundamental frequency, but we need a pattern that is considerably less smooth. I think you can probably guess what we need to do. Yes, it's time for some Fourier synthesis.

To synthesize the stripe pattern, we'll need to add more frequency components into the mix. By the time we've included all harmonics up to the seventeenth (as shown in the inset on the right in the illustration on the previous page), the edges have sharpened up dramatically and the pattern flips from black to white much more quickly. The stripes start to look very much like stripes, although some ripple remains in the flat parts and the edges aren't quite as sharp ("vertical") as we'd like. That's because we've only included seventeen harmonics[13] of the infinite number[14] making up this particular pattern.

The spikes in the graph on the previous page show exactly how *much* of each harmonic is needed to produce the pattern. But how do we know the recipe? For one thing, why is it that, in this case, only the odd harmonics are required? And why do we need so much of the first and third harmonics but so very little of the seventeenth? Those questions can only be answered by biting the bullet and doing the maths, I'm afraid. This is what Fourier analysis gives us—the mix of harmonic ingredients that allow us to reconstruct whatever pattern we like. If you want to get a sense of the workings of Fourier maths, take a look at the appendix.

---

[13] Well, nine actually: the even-numbered harmonics that make up a square wave have zero amplitude. We don't need them.

[14] Yes, you read that correctly. And even with an infinite number we still won't represent this function *perfectly* enough for some mathematicians. (Those of you who are more mathematically inclined should check out the Wiki entry for "Gibbs phenomenon.") We physicists tend to be a little more relaxed about this sort of thing, however. The Fourier series may be an approximation but it's a damn good approximation, and in a world of spherical cows and frictionless surfaces is more than good enough.

As we add harmonics, the pattern looks more and more like the Stryper stripes; the edges sharpen up and become increasingly crisp. Or, more quantum mechanically, to be less uncertain about the positions of the edges, we need a broader range of spatial frequencies. This is again just a visual analog of the maths of music and physics of sound we've covered previously. But this time the most appropriate comparison is with guitar chugs. Remember that in the previous chapter we needed to add frequency components to the mix to produce a kerrchunk sound that had a short duration? This is precisely what we're seeing with the stripes: a broader range of frequency components is necessary to capture what's happening on short *length* scales, rather than short *time* scales. Nonetheless, the physics and maths are identical, no matter the context. The song remains the same . . .

Narrow in time, wide in frequency; narrow in space, wide in (spatial) frequency.

We need to include high (spatial) frequency components to get a high-definition image. This is because it's those higher frequency components, with their short wavelengths, that vary most quickly in space. The more rapidly they vary, the more accurately—sharply—they can define an edge. Or, in language that is a little more quantum flavored: without those higher harmonics, we're much more uncertain about the locations of the edges. They're blurred or smoothed out.

We've arrived at the uncertainty principle again. But now it's in a context that's rather more familiar to the quantum physicist.[15] The first time the vast majority of physics students encounter the uncertainty principle, it's (unfortunately) not in the context of frequency and time, as we covered in previous chapters. Heisenberg formulated the principle in response to the maddening refusal of particles to be pinned down in two domains at once, and physics students consider it in this context, focusing on how uncertainties in position and momentum of a quantum

---

[15] Well, *okay*, maybe not the Stryper aspect. I mean the real space vs. reciprocal space concept.

particle trade off against each other: a higher level of uncertainty about where we can locate a particle in space means a lower level of uncertainty about its momentum, and vice versa.

The explanation for this trade-off in uncertainty that is often given to students is the *gedankenexperiment* known as Heisenberg's microscope. Heisenberg rationalized the uncertainty principle as follows. He argued that to measure the position of an electron, we'd have to use photons with a very short wavelength.[16] Gamma rays have very short wavelengths indeed, and so Heisenberg imagined having a super-resolution gamma ray microscope. Gamma ray photons were shone onto electrons—in Heisenberg's world of imagination—to get a measurement of their position. The problem, however, is that the photons disturb the trajectory of the electrons—they affect their momentum.

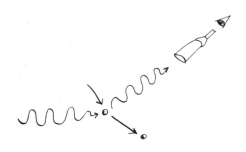

Heisenberg assumed that the origin of the uncertainty principle was the disturbance introduced by the measurement: the photons scatter off the electrons, affecting their trajectory and therefore leading to an uncertainty in just how fast, and in which direction, the particles are actually traveling. As we've said, this interpretation—where the observer or measurer is the fundamental source of the uncertainty—is buried deep in the public understanding of Heisenberg's uncertainty principle. Here, for example, is how HowStuffWorks.com explains the uncertainty principle:

---

[16] Scanning probes were many, many decades away at the time Heisenberg was cogitating about all of this.

Werner Heisenberg, a German physicist, determined that our observa-
tions have an effect on the behavior of quanta . . . To know the velocity
of a quark we must measure it, and to measure it, we are forced to affect
it. The same goes for observing an object's position.

Indeed, the observer effect in quantum mechanics remains one of the more troublesome and controversial aspects of the entire theoretical framework. The "traditional" approach to quantum mechanics,[17] known as the Copenhagen interpretation, describes the measurement process as a collapse of the wave function. That nebulous probability wave collapses down to a specific value when a measurement is made. For those of you familiar with Matt Groening's *Futurama,* you might remember the episode where Professor Farnsworth and the crew of the Planet Express have a day out at the races. After Farnsworth's horse loses in a "quantum finish," he protests: "No fair! You changed the outcome by measuring it."

But Farnsworth is not referring to the uncertainty principle. The observer effect is distinct from Heisenberg's famous principle. While Heisenberg was correct both about the uncertainty principle and the existence of the observer effect, he was mistaken in thinking that one was solely the result of the other. The uncertainty for a wave is just . . . there, whether an observer is or not. Uncertainty is an essential natural feature of waves: a leopard has its spots, Stryper has their stripes, and waves have the uncertainty principle.

Quantum physics is, at its core, a theory of waves; all the evidence points to the inescapable conclusion that matter is imbued with wave-like characteristics. Accepting that means taking into account all we know about the properties of waves. And a key property of any wave, be it at the nanoscopic quantum level or in the not-so-nano world of metal riffs and iconography, is that frequency and time—or spatial frequency

---

[17] Quantum mechanics comes in many different flavors, including pilot wave theory, the many-worlds interpretation, Quantum Bayesianism, Quantum Darwinism . . . There are dozens of versions. It's almost like physicists don't have a good, solid grasp of just what's happening down there at the quantum level . . .

and space—are reciprocally related. There's nothing weird, wacky, or quantum about this. It's just how waves behave; it's what waves do.

Keeping Up the Momentum

We've got one last piece of the uncertainty puzzle to put in place. We know that uncertainty in spatial frequency goes up as uncertainty in position goes down: those stripes got sharper as the spectrum got broader. But how does spatial frequency relate to momentum at the quantum level?

To answer that question, it's time to go back to de Broglie. Almost everything we need in order to connect the momentum of a quantum particle with its spatial frequency is embedded in his simple equation, which I'm expressing this time in terms of momentum rather than wavelength:

$$p = h/\lambda$$

To recap: in de Broglie's equation, p represents the momentum of a particle, h is Planck's ubiquitous constant, and λ is the wavelength associated with the particle. (In some quantum circles, this dual wave-particle entity is known as a wavicle.) When it comes to the relationship of those garish Stryper patterns and momentum at the quantum level, de Broglie's seemingly simple equation again tells us a great deal. Note that the wavelength, λ, is in the denominator on the right side. This means that to find the momentum of a particle we need to consider not its wavelength, but the *reciprocal* of its wavelength. And the conversion factor—the "exchange rate," if you will—from spatial frequency to momentum is Planck's constant.[18]

Wavelength is measured in units of meters: kilometers for radio waves, centimeters for microwaves, and, as you know, nanometers (and

---

[18] And, as we saw in Chapter 6, Planck's constant is also the conversion factor between energy and frequency: $E = hf$.

below) for quantum particles. The reciprocal of wavelength has units of 1/meters, i.e., m^{-1}. It's a measure of how many full wavelengths we have per meter: a spatial frequency. It's *exactly* the same idea as measuring the number of Stryper's stripes per meter.

Except, of course, that the origin of the wave in the quantum case is not quite as conceptually simple to grasp as those spandex patterns. As we've seen, a quantum wave relates to the probability of finding the particle at certain locations in time and space—hardly the easiest idea to get one's head around. But, like all good physicists who hold to the "Shut up and calculate" maxim,[19] let's put aside the thorny philosophical ramifications. That may seem like rather a large cop-out, but debates still rage among physicists (and philosophers) as to just what quantum waves actually represent: What does it all mean? Even in—*especially* in—undergraduate physics courses, interpretational issues are often swept under the carpet. As Lee Smolin discussed in his 2006 book, *The Trouble with Physics,* as a discipline we need to get back to the profound debates that characterized the early days of quantum mechanics and rely a little less on the "Well, the maths works out" argument. In some cases, though, some selective carpet sweeping is the only way to clear a path from one point to another.

We may not fully understand just what's waving at the quantum level, but the message from de Broglie's equation is clear: the momentum, p,

---

[19] The "Shut up and calculate" approach to quantum mechanics is often attributed to Richard Feynman. It indeed sounds very much like the type of thing he'd say. But it appears that the quote has been misattributed. (Misattribution and misrepresentation on the internet? Surely not.) It looks likely that the origin of the famous dictum was, in fact, the solid-state physicist David Mermin. As Mermin writes in an online article about the roots of the phrase: "In short, I suspect that it is only Feynman's habitual irreverence that has linked him in the minds of many to the phrase 'shut up and calculate.' Who else among the high and mighty—and Merton has taught us that it is only among the high and mighty that people tend to look—could have said it? Albert Einstein? Don't be silly. Erwin Schrödinger? Of course not. Niels Bohr? Don't make me laugh. None of them besides Feynman could have said it. Does that mean that Feynman said it? No!" (Mermin, David N. "Could Feynman have said this?" *Physics Today* 57, no. 5 (2004): 10.)

of a particle, is directly related to the spatial frequency of the quantum wave. Smaller, higher energy wavelengths come with higher momentum. In other words, the more the wave wiggles, the punchier the particle gets.

Color Our World Flattened

It's about time we left those spandex trousers behind. They've helped, I hope, to clarify the concept of spatial frequency but, like some other aspects of popular (and not so popular) culture from that bygone era, spandex really shouldn't be seen outside the reunion tours of '80s hair metal and NWOBHM bands. So, having considered one-dimensional spatial patterns in spandex, let's wisely move on before considering dimensions and dimensionality. We've already seen that symmetry is key to many phenomena in physics; dimensionality is a closely related, and equally important, property.

If we could reduce our three-dimensional world[20] to two dimensions, we'd end up living in a reality for which Edwin Abbott Abbott, a schoolmaster writing at the end of the nineteenth century, coined the name "Flatland." In *Flatland: A Romance of Many Dimensions*, published in 1884, Abbott imagined a two-dimensional world where men were polygons and women were line segments, to satirize the hierarchy of Victorian culture. In the course of the novel, both Lineland, a one-dimensional world, and Spaceland, a 3-D universe, are also visited and explored.

A great deal of the science carried out by quantum physicists over the years has involved spending a lot of time exploring the scientific equivalent of Lineland and Flatland. Electrons can be confined in one, two, three—and, indeed, zero[21]—dimensions by clever chemistry, materials engineering, and experimental design.

---

[20] I'm talking just about dimensions in space. Let's leave time out of this for now. I mentioned before that there are many, many different interpretations of quantum mechanics. That's true in spades for the physics of time.
[21] *Zero* dimensions? Yes. Bear with me.

We've seen how one-dimensional (the nanostring) and two-dimensional (the quantum corral) nanostructures can be formed by positioning individual atoms using scanning probe microscopes, but that's just scratching the surface of what's possible. Chemists can synthesize gazillions of nanostructures in standard beakers using a process known as self-assembly, exploiting the natural attractive forces between atoms and molecules. Similarly, pieces of silicon can be patterned in myriad ways on exceptionally small scales to confine electrons to one-dimensional channels.

Since the discovery of graphene in 2004 (by Andre Geim and Sergei Novoselov at the University of Manchester), Flatland is now a place regularly visited in physics experiments. Graphene is a single atomic layer of graphite. Graphite, in turn, is a form of carbon that you'll have encountered many times (in the form of pencil "lead"), where the atoms are bonded together to form a chicken-wire mesh; one mesh then stacks on top of another, over and over, to form the crystal.

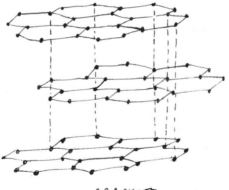

GRAPHITE

Novoselov and Geim's first experiments on graphene are a wonderful example of a Heath Robinson approach to science really paying off. Novoselov found that simply by peeling off layers of graphite, using nothing more than ordinary adhesive tape, he could eventually isolate a single atomic layer of carbon: a truly two-dimensional solid. This led to an explosion of interest and research in graphene, and the pair were awarded the Nobel Prize in Physics shortly after the discovery of the material. At the time of this writing, a wide range of 2-D solids—made of layered materials like graphite, including molybdenum disulfide, boron nitride, and tantalum diselenide—are being studied, their confinement of electrons to a single plane offering the kind of unique experimental conditions that lead to entirely new insights into the way the universe works.

The natural end point of reducing dimensions is not, as you might expect, a 1-D string. Rather, it's *zero-dimensional*: an object in which electrons (or other subatomic particles) are confined in all three dimensions. The archetypal example of this extreme of lowered dimensionality is the quantum dot: a tiny chunk of matter comprising a few tens to a few thousands of atoms, where electrons are confined not to a plane (as in graphene), or to a line (as in a nanostring), but to a very small region of space that is rather more akin to a point—a nanoscale point.[22] Hence the term "quantum dot." This nanoscale chunk of matter is also known as an artificial atom. By controlling the size and shape of the nanoparticle, it's possible to tune the energies of its electrons. We could, in principle, build up an artificial periodic table from just a single material. But that's a (lengthy) story for a different book.

Living on a Razor's Edge . . .[23]

You are no doubt eager to explore the links between electrons confined to Flatland and those spiky logos beloved of metal bands. The members

---

[22] Hang on, doesn't something seem a little amiss here? A flat plane of atoms is two-dimensional, a nanowire that is one atom thick is one-dimensional, so how in the name of all that's unholy is a chunk of material comprising thousands of atoms—a chunk of matter—*zero* dimensional? Isn't it *three* dimensional? Well, yes. And no. (Isn't physics great?) In terms of its atomic structure a nanoparticle is indeed three dimensional. But what's important for the electrons *inside* the particle is how they're confined. In a 2-D system like graphene, the electrons are confined to a plane (but free to move about that plane); in a 1-D structure, they're confined to move to and fro along a line; and in a 0-D structure—like a nanoparticle—they're confined from every dimension and not free to move at all. Hence, zero dimensionality.

[23] From Iron Maiden. "The Evil That Men Do." *Seventh Son of a Seventh Son.* Musicland Studios, 1988.

of Metallizer, following faithfully in the footsteps of their idols, Maiden, Priest, and Metallica, have, of course, chosen a suitably spiky logo for their band. You may have spotted this occasionally elsewhere in the book (emblazoned on the side of their tour bus, for example), but let's see it in all its glory:

Metallizer's logo is a two-dimensional pattern. Thus far, we've seen frequency spectra for 1-D patterns in both time and space, but what does the (spatial) frequency spectrum for a 2-D pattern look like? Well, here it is:

That's a map of the reciprocal space representation of Metallizer's logo. Now that may look, and sound, complicated, but, like many aspects of Fourier analysis, it's not as difficult to interpret as might first seem to be the case. When we had the cross-section of the Stryper stripes, we had one dimension; the pattern had a certain repeat distance along its length. One dimension in space means that we also had a one-dimensional Fourier spectrum—1-D reciprocal space—that showed us the spatial frequencies of the waves required to generate that one-dimensional (or "unidirectional") pattern.

For the Metallizer logo, we have two dimensions in (real) space. This means that there are also two dimensions in reciprocal space. These relate to the spatial frequencies along the length and breadth, i.e., the *x* and *y* directions, of the logo.[24] Here's how it works for the x-direction:

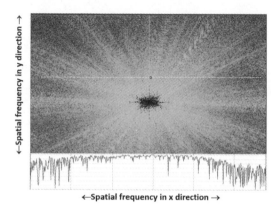

←Spatial frequency in x direction →

We've cut through the image to produce a 1-D slice. The 2-D image is simply a stack of those 1-D slices.

To bring this home, let's see what happens to the Metallizer logo when we filter out certain regions of the reciprocal space map. Here's the result when we remove the higher frequencies:

---

[24] Those of you with some experience in Fourier processing of images will realize that I'm only telling half the story here. What I'm very deliberately leaving out is the wonderfully thorny concept of negative frequencies. I've left that particular aspect of Fourier analysis to the appendix.

For those of you who've used Photoshop or other similar image-processing applications, what we've done above is perform a low-pass filter on the image: we've blocked high spatial frequencies, but retained low spatial frequencies. (Hence, a "low-pass" filter.) We already know what the result is going be from the Stryper spandex analysis: the image has lost definition. The Metallizer logo is blurry and has lost its spikiness because the high spatial frequency components are needed for sharp edges. Having low-pass filtered the image, all we have left are the waves that very slowly undulate—they don't change on short enough length scales to define an edge. Similarly, Party Cannon's bubbly cartoon stands in stark contrast to the extreme death metal spikiness of the Cattle Decapitation logo largely because the latter contains much higher spatial frequencies that give it the edge.[25]

What happens if we filter at the other end of the frequency scale? Well, this is what we get:

This time we've removed the low spatial frequencies and kept the high frequencies. The effect on the logo shouldn't come as a surprise this time, either: we've suppressed the waves that vary slowly across the image so that the rapidly varying waves come to the fore. This means that sharp edges are accentuated, because they're defined by the higher spatial frequencies. Think back to those spandex trousers (sorry), and

---

[25] Purely from a physics perspective, you understand. I, for one, much prefer Party Cannon's take on heavy metal iconography for all of the reasons discussed earlier in the chapter.

remember that only when we added in the higher harmonics (that is, the higher spatial frequencies) did we reproduce the sharp edges of the stripes. It's the same principle here.

Just how does this deconstruction and filtering of the Metallizer logo connect with the quantum world? We saw a few chapters back that nanoscientists can manipulate atoms to spell out almost arbitrary patterns and words. There was a trend in the early days of atomic manipulation to arrange atoms to spell out the initials of the university/ institute where the work was based. (Indeed, you may recall that the first example of atomic positioning involved spelling out the IBM logo.) But is there a limit to how sharp we can make a quantum pattern?

It turns out that, remarkably, nature imposes its own bounds on spatial frequency at the quantum level; fundamental physics and chemistry set natural frequency limits. Arbitrarily high spatial frequencies can't be reached because that would require too much energy for the electrons.[26] This means that electron waves are naturally low-pass filtered. And that, in turn, means that there's a limit to how edgy we can get.

. . . Balancing on a Ledge[27]

Scanning probe microscopists who map and explore electron waves on the surfaces of metal crystals are very aware of the limitation nature sets on edginess. The nanoscopic landscape shown on the next page is a scanning tunneling microscope image of atomic steps and some defects on a silver surface. Atomic steps like the ones seen in the image are extremely common in the nanoworld. Although we nanoscientists can prepare surfaces that are atomically flat over large areas (many thousands of atoms, in the best case), we can't beat the laws of thermodynamics. Defects are an inevitability, and step edges are among the most inevitable.

---

[26] The higher the frequency, the more energetic the quantum wave.
[27] And ditto . . . "The Evil That Men Do."

Image by Adam Sweetman, now at University of Leeds.

An atomic step, where a flat terrace of atoms is broken up by a ledge, represents the sharpest edge we can achieve in any material: the smallest building block of our material world is the atom, so the edge of a plane of atoms is about as edgy as it gets.

Those ripples decaying away from the edges in the image above? They're electron waves. Just like the ripples inside the circular quantum corral a few chapters back, we're seeing electrons scattered from the edge. This time the structure that is the source of the wave pattern is a natural one—it formed during the preparation of the crystal surface[28]—rather than an artificial nanostructure created by sliding atoms across the surface one at a time. But ultimately the same physics are at work: the wave characteristics of the surface electrons are plain to see in both cases.

What's fascinating about the step-edge ripples is that they represent an instance of natural low-pass filtering in action. The electron waves,

---

[28] This involves bombarding the surface, in a vacuum, with charged atoms (ions) from a gun. (It's literally called an ion gun.) Doing so sputters off any crud and contamination on the sample, which is then heated up to remove the damage caused by hammering the surface with a stream of ionic "bullets." After this process is repeated a few times—or in the worst cases, seemingly a few hundred times—an atomically flat surface with a low number of steps is attained.

due to the physical and chemical properties of the silver surface, have a maximum spatial frequency they can reach. Frequencies above that limit simply can't exist. At the atomic level, physics does the Photoshopping. But we also see exactly this effect in the low-pass filtered version of the Metallizer logo. Look carefully at the edges of the logo— you'll see residual ripples there. It's the same mathematics in action: to define a sharp edge, we need high spatial frequencies. Remove those higher Fourier components and not only do edges become blurred, but we don't have the correct mix of waves to ensure that, away from the edge, everything flattens out in just the right way.

Those ripples on the metal surface are known as Friedel oscillations, after the French physicist Jacques Friedel, whose PhD thesis research focused on how electrons rearrange themselves around defects and charges. In addition to the step-edge waviness, you can spot that there are a few contaminant atoms (or molecules) in the STM image on the previous page where the electron waves form rings around the defects. It was exactly this effect that Friedel predicted and described mathematically in his work at the University of Bristol from 1949 to 1952.[29] The rings arise for the same reason that we see ripples at the step edge. The electrons want to "shield" the rest of the surface from the effect of the defect and so crowd around it.[30] They can't shield it perfectly, however, because of the fundamental spatial frequency limit that constrains the highest Fourier components that are possible. The lack of those higher frequency components means that it's not possible to define a sharp edge: ripples are the best that nature allows. The difference in symmetry—a round atom

---

[29] Friedel's PhD thesis advisor was Nevill Mott—a huge name in the field of solid-state/condensed-matter physics. Mott won the Nobel Prize in Physics in 1977, along with Philip Anderson and John Hasbrouck van Vleck, for his work on magnetism and the behavior of electrons in disordered materials.

[30] There's another example of the classic physicist's anthropomorphism to which I referred a while ago. The electrons of course don't "want" to do anything. The physics is such that the electrons are subject to forces with the net result that, to minimize the energy, the influence of the defect is "screened"/shielded from the surrounding material.

rather than a straight edge—results in concentric rings of electron waves rather than lines of ripples.

To supplement all the conceptual metal-quantum links we've explored, I worked with my friend and colleague Adam Sweetman, a talented probe microscopist and major metal fan, to create something tangible: constructing a nanoscopic realization of the universal metal symbol, \m/, by positioning heavy metal atoms (silver) one at a time.[31] The picture below is the metal nanostructure, fabricated and imaged using an STM. The ripples—the Friedel oscillations—surrounding the atomic horns once again arise from the lack of electron waves having high enough spatial frequencies to faithfully reproduce the edge definition of the \m/ sign.

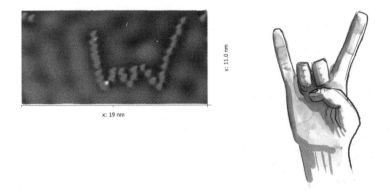

As he developed his mathematics, Fourier had no idea that his brilliant insights into the anatomy of patterns were woven into the very fabric of reality: the mind-bending, revolutionary discovery that matter had a dual wave-particle character was over a century away. And it's not just the material, physical, "IRL" world that reflects Fourier's mathematical

---

[31] A commercial scanning probe microscope system (Scienta Omicron) operating at a temperature of 5 K (i.e., five degrees above absolute zero) and in a vacuum of 10^{-11} mbar was used to build the nanostructure. (The millibar unit represents one thousandth of atmospheric pressure: 1 bar.) The tip of the microscope was used to position Ag atoms that had previously been "dug" out of the surface (also using the probe).

framework; that framework is also the scaffolding on which virtual reality—the images we see and the sounds we hear each day on the internet—is built.

JPEG image compression is fundamentally based on Fourier filtering—discarding higher spatial frequency components reduces the file size. Every compressed JPEG image that's ever been created—all those playful cats, satirical memes, and satirical playful cat memes—has exploited Fourier's mathematics to isolate and remove the higher frequency components. The removal of higher spatial frequencies can, however, be taken to unnecessary extremes. You may have seen the "Needs More JPEG" meme (or the "edgier" version, "Needs Moar JPEG") that did the rounds on the internet. (If not, Google is your friend here.) That meme is nothing more than an example of extreme Fourier filtering in action. Apply that type of compression to the death metal logos that opened the chapter and any vestiges of intelligibility would be washed out entirely.

Given that filtering and removal of spatial frequencies can be used to reduce the size of image files, you might be wondering whether a similar strategy can be applied to audio files. The answer is aptly provided by the title of a Gizmodo article I stumbled across recently: "Digital Music Couldn't Exist Without the Fourier Transform."[32] MP3 files are based on a compression algorithm concept very similar to that used for JPEGs: Fourier analysis is used to locate and isolate—or, in metal terminology, seek and destroy—high frequency components in the audio signal. It's the same old story, same old song and dance (. . . my

---

[32] Again, Google it. But be aware that the article in question doesn't get the description of the Fourier analysis of musical notes right. Like a typical academic, I'll leave it as an exercise to the reader, based on what we've discussed in earlier chapters, to find where the explanation goes awry in that article. (If you've not encountered Gizmodo before, here's how they describe themselves: "Gizmodo UK isn't your average tech site. Sure, we love technology and gadgets, but like any good nerdlinger, we're also interested in science and smart design.")

friends). It doesn't matter whether we're talking about time or space (or both), breaking a pattern down into its component frequencies is an immensely powerful tool for the analysis, storage, and manipulation of everything from "Symptom of the Universe" to truly universal signals.

Chapter 9

INTO THE VOID

Past the stars in fields of ancient void
Through the shields of darkness where they find

—from Black Sabbath's "Into the Void"[1]

---

[1] Written by Iommi/Ward/Butler/Osbourne. From *Master of Reality*, Warner Bros. Records (1971).

Crowd-sourcing data analysis is a canny strategy for scientists who need to hunt through and/or classify vast quantities of information. Zooniverse is a pioneering example of just how powerful citizen science can be: volunteers across the world complete research tasks in a range of disciplines including astronomy, climate science, and biology. At the time of this writing, the Zooniverse community is well over a million strong, and the analysis carried out by self-styled "Zooites" has led to the publication of more than one hundred research papers.

Galaxy Zoo is a core Zooniverse project. It involves the classification of the morphology of galaxies from images taken by the Hubble Space Telescope and the Sloan Digital Sky Survey. There's no comparable project within Zooniverse to analyze matter on rather smaller length scales, i.e., at the nano/quantum level, but nanoscientists at the London Centre for Nanotechnology led by Cyrus Hirjibehedin established a nanoscale version of Galaxy Zoo called Feynman's Flowers—a citizen science project for the classification and analysis of single magnetic molecules imaged with an STM.

Before Zooniverse, there was a less interactive, but nonetheless extremely powerful, approach to citizen science: SETI@home. This is a distributed computing spin-off that makes use of spare processor time/power on personal computers (and associated devices) across the world to analyze signals in a search for extraterrestrial intelligence (SETI). Those who contribute to SETI@home download a piece of code that runs either as a screensaver or as a background process, continuously analyzing radio telescope data from the Arecibo Observatory in Puerto Rico.[2] Radio telescopes are used to map and analyze the radio frequency

---

[2] The Arecibo Observatory, which, coincidentally, features in the 1997 movie *Contact* (based on the 1985 Carl Sagan novel), sustained damage due to

band of the electromagnetic spectrum; it's the same principle as a standard telescope except that instead of capturing light waves, much lower frequency radio waves are detected.

At this point in this book, you'll not be surprised in the slightest to hear that Fourier analysis is at the heart of the SETI@home data analysis strategy. The image below is a screenshot of the SETI@home data-crunching application hard at work. Note the words "Fast Fourier Transform"[3] in the top left-hand corner: the radio telescope data is Fourier transformed in order to hunt for specific signals in the jumble of radio frequency (RF) noise.

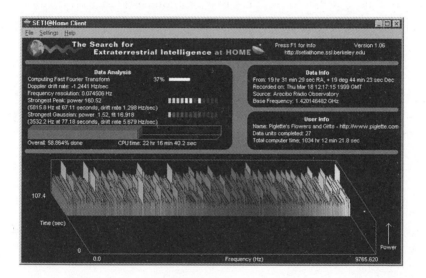

SETI@home focuses on detecting frequency spikes, also known as narrow-band signals, in the otherwise random noise that's collected by the radio telescope. The key idea here is that communication from our

---

Hurricane Maria in 2017 and its fate hung in the balance for a number of months. Fortunately, the US National Science Foundation agreed in November 2017 to continue to fund the facility.

[3] The Fast Fourier Transform (FFT) is the most common approach to computationally calculating Fourier spectra. As the name suggests, it's an algorithm—or, more correctly, a class of algorithms—that can compute Fourier spectra very quickly indeed.

small, furry counterparts on a planet orbiting Alpha Centauri[4] will be based on their broadcasting signals at specific frequencies, or within bands of frequencies, just like our communications on Earth.

But, remarkably, even background "noise," or what at first seems to be just hiss and fuzz, can contain not only important, but groundbreaking, information, information that revolutionizes our understanding of the universe. Much like black metal may appear to be pure aural noise to those whose ears have not been trained—through repeated exposure to blast beats, grinding guitars, and the most guttural of death grunts— to appreciate the musicality of 275 bpm double bass drumming, it's more than possible for a signal that contains essential information to be dismissed by scientists as originating from nothing more than, um, pigeon droppings. And, indeed, this is exactly how the characteristic signal of the birth of our universe was first pooh-poohed.[5]

Back in 1964, physicists Arno Penzias and Robert Wilson were working at Bell Labs on a highly sensitive radio telescope. Despite cooling the equipment down to liquid helium temperature (which, you may recall from a previous chapter, is a shockingly chilly −269°C),

---

[4] Given the discovery in 2016 of an exoplanet orbiting the red dwarf star Proxima Centauri, those communications might be more likely to originate from the system of the smaller Proxima, rather than its bigger siblings, Alpha Centauri A and Alpha Centauri B. But then my allusion to Mr. Adams's work would be lost. (www.clivebanks.co.uk/THHGTTG/THHGTTGradio3.htm.)

[5] Healthy skepticism is the bedrock of science. To quote Feynman yet again, "The first principle is that you must not fool yourself. And you are the easiest person to fool." Much better to first dismiss a signal as arising from pigeon droppings; clean and adjust the equipment; re-re-repeat your measurements; sleep on it; repeat the measurements again; fret and worry; dismantle, rebuild, and recalibrate the instrument; and then, possibly, tentatively, suggest that there might, potentially, perhaps be a signal there. It's rather less scientific to simply jump to conclusions, particularly if you subsequently have to do a volte-face and admit that your "groundbreaking" result was nothing more than feces-induced fluctuations. Unfortunately, in the rat race that is science today, Feynman's dictum has mutated into, "The first principle is that you must get your paper past the journal referees and the editors." And in that pursuit, rigor is too often trumped by newsworthiness.

accounting for terrestrial radio and radar interference, and giving the telescope a good wash-down after removing the aforementioned droppings (and protecting the 'scope from their avian origin), they were still left with a persistent noise in the signals they detected. It was there day and night, all across the sky. An all-pervasive background hum.

It turns out that what Penzias and Wilson had discovered was the afterglow of the Big Bang. They had captured the oldest light in the universe, the "relic radiation," as it is known, that arose a little less than 400,000 years following the beyond-tumultuous birth of everything. (A cosmic blink of an eye in the context of our universe's 13.8 billion-year history.) At that stage in its infancy, the universe had cooled sufficiently so that protons could hold on to electrons, forming hydrogen atoms. Prior to that point in time, all matter formed a dense soup of subatomic particles through which light couldn't pass. But at the tender age of 380,000 years old, the universe's atoms had formed, and photons of light could finally make their way through the fog.

In the 13.4 billion years that have passed since the universe first became transparent to those photons, it's expanded. A lot. And with that expansion, the wavelength of the photons has in turn increased: they've been "redshifted" down the electromagnetic spectrum to the microwave range. The wavelength of the light stemming from the Big Bang is no longer in the range associated with visible light—the colors of the rainbow run from about 400 nm to 700 nm—but instead is about a millimeter. Penzias and Wilson were the first to measure these redshifted photons, now known as the cosmic microwave background (CMB).

All matter emits thermal radiation of some form. You may be aware of the connection between infrared radiation and heat energy, but the temperature/radiation relationship covers a much broader spectrum. A good example is the humble incandescent lightbulb: the visible light given out is a form of thermal radiation. Change the temperature of the filament in the bulb and the peak wavelength of the light emitted will shift. In the case of the cosmic microwave background, the photons have much lower energy than those of visible light: the CMB radiation has an average temperature of 2.7 K (about –270 degrees Celsius, a couple of degrees cooler than the liquid helium Penzias and Wilson used to cool their telescope).

If detecting the primordial photons wasn't cool enough by itself, Penzias and Wilson's discovery had huge implications—though they didn't know it at the time—for our understanding of the *sound* of the early universe. When physicists first developed techniques (derived from an analysis of the CMB) to determine how sound waves traveled across the universe in the early years of its life, they were gobsmacked to find that, just like a guitar string, there were certain preferred, quantized frequencies! Even at the dawn of time and space, Fourier's mathematics resonate.

A Cosmic Cadence[6]

Although the average temperature of the CMB radiation is about 2.7 K, there are tiny fluctuations in that temperature as we look across the sky (i.e., across the universe). The map on the next page, taken by a European Space Agency (ESA) space observatory fittingly named in Planck's honor, shows those fluctuations in spectacular detail, where the shades of gray represent temperature variations around the mean value of 2.7 K in the hundreds of microKelvin (sometimes written as μK). These are exceedingly small differences in temperature—about one part in ten thousand.

---

[6] This section has been informed and inspired to quite some extent by Mark Whittle's wonderfully clear and engaging description of Big Bang acoustics. Professor Whittle is an astronomer at the University of Virginia, Charlottesville, who has written at length, and in a very accessible style, about sound waves in the early universe. Peter Coles has also mused over the sound of the Big Bang— telescoper.wordpress.com/2009/04/26/how-loud-was-the-big-bang/—citing John Cramer's work at the University of Washington, Seattle.

Those teensy temperature variations contain a wealth of information. Remarkably, the fluctuations in the microwave background are an imprint of the variations in the pressure of the primordial soup[7] all those billions of years ago. And just like pressure fluctuations in the air around us give rise to sounds ranging from the most delicate of whispers to the "everything louder than everything else" of a Motörhead, Metallica, or Maiden at full tilt, that blotchy CMB pattern is the signature of universal sound waves.

I'll admit that the idea of a universe-spanning sound wave seems rather bizarre. After all, as the tagline for *Alien* told us all those years ago: "In space no one can hear you scream." That's a sound and scientifically accurate piece of advice—sound waves indeed can't travel through a vacuum because there aren't enough molecules of matter to transport the wave. Remove the atmosphere, and the wave energy can't be transmitted. In other words, it doesn't matter if Metallizer have pumped their amps to eleven and their backline is capable of rending the earth asunder—without an atmosphere, their music isn't going anywhere.

---

[7] Alternatively, we could consider these fluctuations as variations in density. We'll stick with the pressure interpretation. In either case, our current understanding is that the source of the variations in density and pressure that gave rise to the CMB pattern was the presence of quantum fluctuations in the moments directly following the Big Bang. Those tiny quantum fluctuations were then magnified massively via a process known as inflation. But that's a whole other, ever-expanding, story.

At the birth of the universe, however, there *was* an atmosphere. In fact, you could say that there was nothing but atmosphere. The primordial soup was everywhere and everything. It's not like the universe today, where space has expanded dramatically and matter has clumped together to form stars, galaxies, and planets with vast swaths of vacuous nothingness in between. The universal sound waves etched into the CMB formed when the universe was so small that this nothingness didn't exist—matter was everything. Nothing else mattered.

How loud were those sound waves at the dawn of time and space? Mark Whittle's analysis of Big Bang acoustics[8] shows that, in perhaps the most seminal connection between metal and quantum physics we can imagine, the fluctuations were at the level of 110 dB. That's pretty much the sound pressure level—the loudness—of a typical metal gig. To understand how Whittle determined that figure, we need to take just a moment to examine what a decibel represents. (I did promise all the way back in Chapter 3 that we'd eventually get round to an explanation of the decibel. Apologies for taking so long.)

Although it's most commonly used in the context of loudness, the decibel is a rather more general—perhaps I should say universal—method of comparing two quantities that's used all across science and engineering. For our purpose, we don't need to worry about those other applications. We just need to know how the decibel characterizes how loud something gets. (And, as you may have heard, "If it's too loud, you're too old."[9])

---

[8] Whittle, Mark. "Big Bang Acoustics: Sounds from the Newborn Universe." Department of Astronomy, University of Virginia (Web page). people.virginia .edu/~dmw8f/BBA_web/bba_home.html.

[9] I long assumed that it was Lemmy who coined this phrase—it's got Mr. Kilmister's acerbic wisdom written all over it—but the aphorism instead appears to have originated with Ted Nugent. This is hardly surprising as Mr. Nugent's rants alone, even in the absence of a microphone, tip the decibel meter into the red. Nugent also once claimed that the sound levels from his amplifiers had resulted in the untimely passing of a passing pigeon. Purely apocryphal, of course, but the image of a pigeon being blown apart by the force of Ted Nugent's decibels has passed into rock legend and helped foster the myth of the self-styled

The "bel" bit of "decibel" was named after Alexander Graham Bell, the inventor of the telephone.[10] A bel—let's give it the symbol B—compares the loudness of two sounds (or signals) like this:

$$B = \log_{10}(P_{\backslash m/}/P_{Sshh})$$

In that equation, $P_{\backslash m/}$ denotes the sound pressure level of our loud signal, whereas P_{Sshh} represents the reference level. For sound, that reference level is the pressure associated with sound at the lowest limit of human audibility.[11] That level, the smallest pressure variation that results in something we can hear, is conventionally taken to be about 1/10,000,000,000 (or 10^{-10}) of atmospheric pressure. The bel is the logarithm of the ratio of the two pressures. Logarithms are one of those aspects of high school maths that tend to make non-mathematicians run to the hills but, like so many mathematical concepts, they've got a much scarier reputation than they deserve. A logarithm is just the number of times we multiply the base by itself to get the number in question. The base for bels is 10 (which is why the logarithm is written as $\log_{10}$ in the formula above). It's as simple as this:

$$\log_{10}(10) = 1;$$
$$\log_{10}(100) = 2;$$
$$\log_{10}(1,000) = 3;$$
$$\log_{10}(10^6) = 6;$$
$$\text{and } \log_{10}(10^{666}) = 666.$$

Motor City Madman. Ozzy Osbourne, on the other hand, infamously took a rather more direct approach to dealing with our feathered friends . . .

[10] Like so many scientists and engineers who invent or discover something for which they are forever connected in the public imagination, Bell's development of the phone overshadows his achievements in a range of areas. He was also rather perceptive and prescient with regard to technological development. Over a century ago he said the following: "Every town or city has a vast expanse of roof exposed to the sun. There is no reason why we should not use the roofs of our houses to install solar apparatus to catch and store the heat received from the sun," and, "The day will come when the man at the telephone will be able to see the distant person to whom he is speaking."

[11] Yes, we could debate at length just how that is defined.

You've seen this type of approach before—it's effectively just the same as scientific notation. We don't write 1,000,000,000,000,000 because that's too much like hard work and it's very easy to lose track of all of those zeroes. Instead we write that number in the shorthand of scientific notation as 10^{15}. The logarithm of 10^{15} (to the base ten) is simply 15.

So to work out a bel, we just divide the pressure levels and get the logarithm. But due to the size of the numbers, it's often more useful to work in units of decibels (dBs). One bel is ten decibels so our formula in principle just needs one simple fix:

$$dB = 10 \log_{10} (P_{\backslash m\!/}/P_{Sshh})$$

However, it's slightly more complicated than that, because the energy associated with the wave increases proportionally to not just the amplitude of the pressure variation (remember that we're talking about pressure waves here) but to the *square* of the amplitude (don't worry too much about why this is—it's a subtlety with regard to amplitude). So that gives us another factor of two in the equation:

$$dB = 20 \log_{10} (P_{\backslash m\!/}/P_{Sshh})$$

The reason this logarithmic scale is so useful when it comes to sound is that it matches just how our ears perceive sound. They respond logarithmically, not linearly, to changes in sound intensity.[12] The limits of our hearing are quite remarkable: a healthy human's hearing spans a range of about 130 dB. The quietest sound we can hear is about 13 orders of magnitude (one ten-trillionth) smaller in terms of sound pressure level than the noisiest. Manowar claims the title of the loudest band in the world, clocking in at a sound pressure level of 139 dB.

---

[12] I am glossing over a great deal of sophisticated physics, biology, and psychology here. The science of hearing is immensely complicated, and I am presenting just the bare bones. To any acousticians who may be reading— apologies for not doing your multifaceted subject justice.

That's very much pushing the tolerance of human hearing[13]—sustained exposure to sound waves of that intensity can produce serious damage.

So just how many dBs were associated with those primordial pressure variations imprinted in the cosmic microwave background? Before we can work that out, we need to somehow define our reference level, our P_{Shhh}, for the CMB. As mentioned above, for sound waves in air, the convention is to define P_{Shhh} as a fraction of atmospheric pressure of 10^{-10}—taken to be the limit of human hearing. Let's adopt this approach and take our reference level for the CMB to be similarly a factor of 10^{-10} times the ambient pressure of the primordial gas. The pressure variations can be determined from the fluctuations in the CMB and turn out to be between 10^{-4} and 10^{-5}. So that means we have variations that are between 10^6 (10^{-4} divided by 10^{-10}) and 10^5 (10^{-5} divided by 10^{-10}) times larger than the reference level. Plug those numbers into our dB formula on the previous page and we find that the Big Bang was between 100 dB and 120 dB of pure primeval, momentous noise.

Let the implications of those numbers sink in for a moment. What that calculation is telling us is that a Manowar gig—139 dB, remember—was *louder than the Big Bang*!

True metal can drown out the sound of the birth throes of the universe.

There's loud, and then there's *LOUD*. (The guys in Manowar are hardly the most self-effacing in the metal community. This piece of information could well cause their egos to expand at a rate that would dwarf the Big Bang–induced expansion of space-time. Best keep it to ourselves.) Leaving aside the punishing volumes achieved by Manowar, Big Bang levels of 100–120 dB are pretty common at any standard

[13] . . . but then there are those who would argue that no matter what the dB reading might be, Manowar's output is at the limits of human tolerance. (Joke. I know how seriously those loincloth-clad purveyors and protectors of true metal take their art. And how much time they've spent working out.)

metal gig (if any metal gig could really be called standard). Those are still tinnitus-inducing sound pressure levels.

But the Big Bang wasn't only loud, it was heavy. *Very* heavy. Mr. Townsend and his erstwhile Strapping Young Lad colleagues would no doubt be pleased to hear that the origin of the universe wasn't only heavy as a really heavy thing, it was heavy like no sound ever since.

Many metal bands very deliberately down tune to eke as much heaviness out of their riffs as possible. (Down tuning means loosening the strings to produce notes lower than the guitar's standard EADGBE.) Tony Iommi, the riff-meister whose gargantuan sound laid the foundations of metal, regularly tuned down from E to C♯ (*Sabbath Bloody Sabbath* being a key example), an approach that Metallica notably adopted for "The Thing That Should Not Be." "Drop D" tuning, where only the lowest string is down tuned, was a massive part of the Seattle grunge sound. (Although King's X had been using it masterfully for a number of years before the Seattle scene exploded. Their *Out of the Silent Planet* debut combines Sabbath's and Metallica's heaviness with sublime melodies and Beatlesesque harmonies, and a great deal of their signature sound arises from that drop D tuning.) Drop D has since been supplemented with a variety of drop tunings, including the rumbling drop B grind regularly exploited by Slipknot (though they're certainly not the only band to have down-tuned to that extent).

The Big Bang took drop tuning to a new low, however. Those sound waves at the dawn of time had frequencies so low as to make even the most heavily down-tuned metal band sound like Alvin and the Chipmunks after they'd inhaled their fill of helium party balloons (while on acid, with Party Cannon as the house band). It wasn't just sub-bass frequencies that resonated across the universe; these were sub-sub-sub-sub-bass, and below. They weren't even measured in units of Hz. Drop B tune your guitar, and the sound of the lowest string will weigh in at a frequency of about 62 Hz, meaning the wave has a period of 1/62 of a second. The universal waves had a period of between 20,000 to 200,000 *years*. That translates to frequencies of between 10^{-12} Hz and 10^{-13} Hz, a demonically deep subsonic boom that is entirely beyond—or below—

our ken. But just like the tuning fork oscillations we covered back in Chapter 7 that were used to image and manipulate chemical bonds, it's possible to transpose the waves into the audible range. To do so, however, in this case requires a shift in frequency of about *fifty octaves*.

Why were the universal waves so low in frequency? Again, it comes back to that simple dictum: the bigger it gets, the heavier it sounds. Large structures are associated with lower resonant frequencies. It doesn't get much bigger than the universe, so it's not entirely surprising that the frequencies of the sound waves imprinted on the CMB are incomprehensibly, inaudibly low—they're so far outside the range of human experience that they span not just generations or millennia, but a period of time that's much longer than all of recorded history. We humans, metal fans or not, don't often get the opportunity to encounter waves whose wavelength spans the entire universe.

Given that I've switched from talking about period to frequency, you might expect a Fourier transform, or its close relative, to make an appearance sometime soon. And here it is:

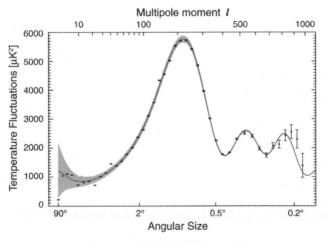

Image from the NASA/WMAP Science Team.

That's the spectrum of the waves that comprise the CMB. There's an awful lot to digest in that graph, but don't let that put you off. Scientists who aren't cosmologists, particle physicists, or astronomers/

astrophysicists—including "squalid-state" physicists like myself[14]—tend to also need a primer on just what the heck that graph actually means. Be assured that I'm not going to bother with the minutiae. We just need to see the big picture. And in this case, it's a very big picture indeed.

Forget about the top axis for the moment—we'll get back to that shortly. The y-axis is relatively straightforward to explain: it's the temperature fluctuations of the CMB. One minor complication is that you might have spotted that the units aren't micro-kelvin (μK). Instead, they're micro-kelvin squared, μK^2. This is because the fluctuations are plotted in terms of the *power* of the waves, which is proportional to amplitude squared. This is a very common practice in Fourier analysis: we plot a spectrum that shows the power of the various frequency components, rather than their amplitude. Power is basically amplitude over time, but the differences between power and amplitude need not concern us for now. All you need to take onboard is that the y-axis is a measure of the strength of the waves contributing to the CMB. And in that sense, it's just like all of the other Fourier spectra we've seen throughout this book.

The x-axes are a little trickier to explain, but ultimately they're just a different way of representing a range of wavelengths or spatial frequencies, exactly like those we've covered earlier in this chapter. The CMB is a map of the temperature fluctuations across the sky. Like *any* pattern—from Stryper's stripes, to the Metallizer logo, to the ripples in the coffee cup and the quantum corral, to every single image that's ever been downloaded, captured on a camera, drawn, or created by

---

[14] Murray Gell-Mann, the brilliant and often irascible particle physicist who not only postulated the existence of quarks but coined the name in the first place, famously described solid-state physics as "squalid-state physics," largely to wind up his non-particle physics colleagues. Some branches of physics have a reputation for attracting physicists who might possibly be described as sometimes lacking a degree of humility. Or at least that's the stereotype. I'm often reminded of this xkcd cartoon: xkcd.com/435/.

some other physical, chemical, and/or biological process—the blotchy, blobby cosmic microwave background is similarly made up of a sum of waves with different spatial frequencies.

What we'd really like to do is graph the power of those waves as a function of their spatial frequency directly, but that's rather tricky. Instead, the power of the waves comprising the CMB pattern is plotted against the angular size of the features. Astronomers do this all the time: angular size is used as a proxy for actual size. Features that don't cover a large part of the sky (i.e., those that are relatively small) subtend a small angle to an observer here on Earth and so are associated with a smaller angular size; the opposite is true for larger features.

That word—*subtend*—tends to garner confused looks at first, but it's simpler than it sounds. We can get a good handle on its meaning via a simple hands-on—sorry, make that hands-off—demonstration. Hold your hand close to your face, say six to nine centimeters away.[15] Now move your line of sight from your fingertips to your wrist. Your eyes will cover quite a large angle as they move to scan across your hand. This, very roughly speaking, is the angle subtended at your eyes. Now move your hand to arm's length and do the same thing. Your eyes will traverse a smaller angular range. A large object in front of you will subtend a larger angle as compared to an object at a distance, because the object at a distance appears smaller.[16]

In the illustration on the next page, the patch of sky highlighted in a brighter shade (and encircled with the dashed line) covers a smaller region of the sky than the surrounding darker "clump." This means it has a smaller angular width.

---

[15] For those whose first language isn't metric, let's call this two to three inches.

[16] If you're a fan of the superlative comedy series *Father Ted*, a certain classic sketch from Series 2 may spring to mind here. (And if you've not seen *Father Ted*, could I suggest that you put this book down now and binge-watch the three series, wherever you can find them. I'll wait for you to get back.)

CMBR

Remember that small-scale features are necessarily associated with higher spatial frequencies: to define a small structure in a pattern, the waves need to vary on a short length scale and thus they have a relatively high spatial frequency. This explains why the numbers on the bottom axis in the CMB spectrum on page 229 are "backward": they run from high to low, rather than low to high (as is traditionally the case for graphs). Larger values of angular size are equivalent to lower spatial frequencies, and vice versa.

That's the bottom axis explained. The top axis is in one sense "just" another way of representing those spatial frequencies: the numbers on the bottom axis can be converted into those on the top axis using the correct "exchange rate." But what the heck, you might very reasonably ask, does the label on that axis—namely, "Multipole moment l"—mean, and how does it relate to spatial frequencies? This is a challenging question, but once we get to the bottom of it you'll have as good an understanding of the CMB as many professional scientists (regardless of how squalid, or not, their chosen field of research might be).

Indeed, in one sense, this question is what we've been steadily building up to throughout this, and the preceding, chapter. We started with 1-D frequency spectra, then moved to the spectra of 2-D images. And thus far we've put the CMB map in the same class as the Metallizer logo—we've treated it as a two-dimensional pattern. It certainly looks like a 2-D pattern when it's displayed on a page or a screen. But the sky, of course, isn't a two-dimensional plane like that on which the Metallizer logo is printed. From our Earth-bound perspective, the actual CMB pattern is best described as being imprinted on a spherical, i.e., three-dimensional, surface. To close this chapter, we're going to once again return—or should that be rotate?—full circle to the essential concept of symmetry.

Music of the Spheres

All the way back in Chapter 2, we saw the intimate connection between circles and sine waves. Marsha the Mosher spent a great deal of time dashing around circle pits to demonstrate the fundamentally circular nature of harmonic motion, and a drum beater helpfully converted circular motion to a sine function for us. Summing up sines has, throughout the book, allowed us to synthesize and desynthesize everything from the sounds of Fear Factory to the electron density in a nanostructure using Fourier's tried, trusted, and centuries-old techniques.

Now, however, we've moved up a dimension (or two) to consider the rather more realistic 3-D nature of the CMB. And that means we need to graduate from circles to spheres, or, more precisely, spherical harmonics. I fully realize that, just like the terms "delta function" and "Fourier transform," the words "spherical harmonic" may strike fear into the heart of all but the most steely-willed physics student. As ever, however, while the mathematical minutiae can sometimes become overwhelming, the underlying concept is really not so difficult to grasp:

four or five years of physics education and training is certainly not a prerequisite to understanding the spherical harmonic.

By now you know very well that Fourier analysis uses sine curves with different frequencies (and amplitudes and phases) as its basic building blocks.[17] When it comes to three-dimensional structures and patterns, however, it is often much easier and much more elegant to use functions that have the appropriate symmetry. And very often that appropriate symmetry is that of a sphere. Why? Because spherical symmetry is *everywhere*. Planets, for instance, ours included, are spheres (or have a shape that is very close indeed to spherical[18]). Have you ever wondered why they aren't cubes, or pyramids, or cylinders, or discs?[19]

The reason for this is simple. Gravity is what's known as a central force: the force depends only on the center-to-center separation of two objects. The upshot of this is that matter will be attracted inwards toward the center of mass. The force acts radially inwards, as in the sketch on the next page. Or, from an equally valid alternative perspective, matter is attracted toward the center of the earth because this minimizes the gravitational potential energy. (You may remember the mosher who fell to Earth/earth in a previous chapter?) If matter is free to move under the force of gravity, then the net result is this:

---

[17] . . . or if you don't, then this book has been a dismal failure. A failure of *St. Anger*, *Van Halen III*, and *Shark Sandwich* proportions combined.

[18] In the alternative reality documented so hilariously in an extensive series of novels by Terry Pratchett, the laws of physics are rather different. A disc-shaped planet, the Discworld, results. Ian Stewart and Jack Cohen's take on Pratchett's wonderfully imaginative universe, *The Science of Discworld* (First Anchor Books, 2014), is a hugely entertaining analysis.

[19] One of the more depressing aspects of the internet, and there are increasingly many, is that it has unearthed a large community of flat Earth "enthusiasts" (or, somewhat less euphemistically, tinfoil-helmeted, conspiracy-theorist, beyond-paranoid, antiscience zealots) out there. It's difficult to know how to counter this lunacy. The standard mantra of the academic, the teacher, and, once upon a time, the politician, viz. "education, education, education," clearly isn't working.

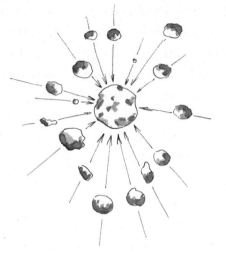

Gravity pulls equally inward in all directions. This means that, as shown in the sketch, the force is spherically symmetric. And matter follows the force of gravity, so a spherical force field means matter will clump together to form spheres. But gravity isn't the only force that acts like this. The electrostatic (sometimes called electrical) force between two charges has the same spherical symmetry: the force between two charged particles depends only on their center-to-center separation. Because of this identical symmetry, there are deep parallels between the behavior of gravity and the electrostatic forces at play within atoms and molecules.

This does not mean that we understand how gravity works at the deepest, quantum mechanical level. The infamous inability to reconcile gravity and quantum mechanics is a continuing source of frustration for physicists. But just like coffee confined within a cup and electrons trapped inside a circular quantum corral each form the same pattern of ripples, the identical spherical symmetry of the gravitational and electrostatic forces means that the same mathematical strategies will often work for both. And those mathematical solutions in turn predict and describe what we see in nature.

In one sense, it's mind-blowing that an identical mathematical framework can describe how the universe behaves on such incomprehensibly different length scales: from universe-spanning ripples in space right down to the physics and chemistry of electrons and other sub-atomic particles. But from the mathematician or physicist's perspective, it's not really so surprising. The gravitational and electrostatic forces might well have entirely different origins, but ultimately they behave identically in one very important way: their force fields have the same symmetry. And that in turn means that the solutions to the mathematic equations that describe those forces are going to have lots in common. It doesn't matter if we're comparing guitar strings with nanowires, or coffee with corrals, or cosmic waves with electronic harmonics; in each case, the core connection is symmetry.

Spherical harmonics capture that three-dimensional symmetry in an elegant mathematical form. Harmonics, in both time and space—well, to be more accurate, in reciprocal time and reciprocal space—have been the bedrock of this book. A harmonic has a frequency that is a multiple of the fundamental frequency; the higher the harmonic, the higher the frequency. In turn, the higher the frequency, the more nodes and antinodes we have in a particular period of time (or, for spatial frequencies, in a particular length or area). Put simply, the higher the harmonic, the more the sine wave function wiggles.

It's very much the same for spherical harmonics. They're three-dimensional, so they're a bit more complicated to get our heads around, but it's just nodes all over again, except with poles/lobes. Here's what a few spherical harmonics look like:

Yes, those look rather perplexing. But we don't need to appreciate any of the subtleties of spherical harmonics—and there are many (just take a look at the Wikipedia entry)—to understand the links with Fourier analysis. (And to finally explain what that "Multipole moment" axis in the CMB spectrum really means.) The family of spherical harmonics on the previous page is organized from top to bottom in terms of frequency. I appreciate that this might be a little difficult to visualize, given that we're dealing with three-dimensional functions, but a helpful way to see the change in frequency is to remember that a higher frequency sine wave will change from positive to negative more often than a low-frequency wave. Like this:

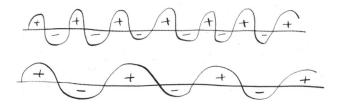

The shading of the spherical harmonics on the previous page shows which part of the function is positive and which is negative. The darker bits are negative. Scan from top to bottom of the spherical harmonic family, and you'll see that the number of positive and negative lobes increases. This means that, just like a sine wave, the frequency is higher for the functions that change more often from plus to minus. And just as we can add up sine waves to produce a 1-D or 2-D pattern, we can add up spherical harmonics to produce patterns in 3-D. Patterns in 3-D are rather more complicated than those in fewer dimensions, so our "ingredients" are rather more complicated as well, but it's exactly the same principle.

In addition to determining frequency by looking at how often the function changes from positive to negative, we can classify the spherical harmonics according to a slightly different, but conceptually very similar, characteristic. Starting again at the harmonic at the top of the hierarchy, note that the function doesn't change its sign, either from top to bottom or side to side: though it is shaded (because it is three dimensional), the sphere is one color. Mathematicians and physicists call this the "monopole" harmonic. Go down a row and there are now two "poles," oriented along the x-, y-, or z-axis as we move from left to right: a dipole harmonic. Go further down the spherical harmonic "family tree," and you'll see that the number of poles increases.

The "Multipole moment *l*" label on the top axis of the CMB spectrum refers precisely to this increase in the number of poles as we add more and more spherical harmonics. As the number of poles increases, the angular width of each pole (or lobe) decreases. A spherical harmonic with a larger number of poles is analogous to a higher frequency sine wave: it accounts for the fine detail in the pattern. This is why the top axis of the CMB spectrum runs from low to high while the bottom axis, as mentioned earlier, has numbers that instead go from high to low—both top and bottom show a left to right progression from, in essence, low to high spatial frequencies. The peaks that we see in the CMB spectrum tell us that not all spherical harmonics—i.e., not all spatial frequencies—are represented equally, otherwise the spectrum would be a flat line. Instead, spherical harmonics with certain numbers of poles are much more common than others, giving rise to the peaks in the spectrum. If this reminds you of our discussion of natural or resonant frequencies on guitar strings, well, it should. The early universe indeed resonated.

The sketches of the spherical harmonics back there might similarly have resonated with those of you who've studied chemistry. An electron in an atom also experiences a spherically symmetric potential due to the nucleus. This has absolutely nothing to do with gravity—electrons and protons couldn't care less about gravity because they're so tiny[20]— but the electrostatic force keeping the electrons from escaping from the

---

[20] The electrostatic force between two electrons is 42 orders of magnitude stronger than the gravitational force between them. I like that the number 42 is riven into the fabric of our universe in this way.

atom is described by a kindred equation. Compare and contrast the formula for the force of gravity, F_{Gr}, between two masses with that for the electrostatic force between two charges, F_{El}:

$$F_{Gr} \propto \frac{m_1 m_2}{r^2}$$

$$F_{El} \propto \frac{q_1 q_2}{r^2}$$

In both cases, the force decreases proportionally[21] to the square of the separation between the masses (i.e. m_1 and m_2) or the charges (q_1 and q_2). The separation is represented by r in both formulae. Both the force of gravity and the electrostatic force are what is known as inverse square laws; the force is proportional to $1/r^2$.

The mathematical similarity of the equations for gravitational and electrostatic force, and their associated identical spherical symmetry, mean that the wave functions for an electron in an atom are the components of the cosmic microwave background pattern writ small—writ very, very small indeed. Electron orbitals are the physical realization of the spherical harmonic functions sketched a few pages back. Even factoring in the physicist's stereotypical stoicism in the face of groundbreaking insights,[22] this is a staggering realization: the universe-spanning CMB pattern is deeply related to the subatomic waves that form on length scales that are unimaginably, incomprehensibly tiny in comparison.

---

[21] I've used a "proportional to" sign ($\propto$) because we don't need to worry about the precise values of the constants that can be included to change that proportionality to an equals sign. All that matters is that the force is proportional to the inverse square of the separation of the masses or the charges.

[22] And this really is nothing more than a stereotype. Physicists are people too, with all the attendant advantages and disadvantages that brings. Or, paraphrasing rather more elegant language courtesy of the Bard: "Hath not a physicist hands, organs, dimensions, senses, affections, passions? Fed with the same food, hurt with the same weapons, subject to the same diseases, healed by the same means, warmed and cooled by the same winter and summer?"

It's mind-blowing that so much information has been extracted from the skies above us and still more astonishing that this information has been used to put our place in the universe into perspective. That a pattern of splodges in the distribution of microwaves could tell us so much about the evolution of our cosmos highlights the ingenuity, imagination, creativity, and sheer bloody-minded tenacity of our species. But then, the scientist in all of us incessantly hunts for patterns, be they in time, space, data, mathematical equations, or the sound of a heavily overdriven guitar.[23]

[23] "When the Uncertainty Principle Goes to Eleven," *passim*.

Chapter 10

IN AND OUT OF PHASE

Standing like a sore thumb out,
Notice your image is different.
. . . I hope you find your niche, someday soon.
Easy to change your phase.

—from Quicksand's "Fazer"[1]

---

[1] Words & Music by Walter Schreifels, Sergio Vega, Thomas Capone & Alan Cage. © Copyright 1993 Bots Entertainment. Universal Music Publishing Limited. All Rights Reserved. International Copyright Secured. Used by permission of Music Sales Limited.

Learning to play guitar certainly ain't what it used to be. I'm of the pre-internet generation, when we couldn't turn to YouTube to find out how to play a riff, a lick, or a full-blown solo by one of our heroes. An era before online tutorials on re-creating every nuance of every aspect of apparently every piece of music ever recorded; a musical Mesozoic, before tablature for everything from "Ace of Spades" to "Zero the Hero" was instantly available (subject to Wi-Fi quality); a halcyon time of yore, when learning Yngwie Malmsteen's greatest hit note-for-note from a precocious nine-year-old on the other side of the world wasn't yet a technologically viable possibility.

Yes, I know I'm sounding like a grumpy old fossil. But there's a lot to be said for the benefits of the old-school approach to learning riffs and solos straight from the source: listen-attempt-fail-repeat. No step-by-step video to follow. No tab at your fingertips. Just a tape recorder playing the same five seconds of a solo over and over until your fingers bleed or your brain screams "no more," whichever comes first.

Does online learning have its advantages? Yes, of course. It's always great when knowledge is shared and, indeed, I've liberally scattered links to online resources and articles throughout this book. But deep learning is much, much more than passive absorption of facts—or, as is now frequently the case, unsubstantiated opinions—from a video stream; it's more than just rote regurgitation of a sequence of steps. I know you know this, but it bears repeating. And it's true whether we're attempting to gain an understanding of the second law of thermodynamics, to solve a particularly thorny differential equation, or to develop the ability to express ourselves via a face-melting guitar solo that says something different than every other virtuoso guitarist we've heard before.

Too often, the key role of confusion in gaining a deep understanding of a subject is overlooked entirely. The mantra in teaching (at any level) is "clarity, clarity, clarity": the prevailing view is that a student is better able to grasp the material when it's laid out in a pristinely clear and coherent step-by-step style (preferably "delivered" by a captivating teacher). The student should come away from a lesson feeling that much smarter.

But the problem is that any substantial leap in my understanding of physics has come not when I've felt smart but when I've felt stupid. Really stupid.

That cognitive "dissonance," that banging-your-head-against-the-wall feeling and the associated wandering around in circles (figuratively, literally, or both) are all key aspects of truly coming to grips with a concept. Sure, a great teacher can be an immense help to the learning process. But I would argue that her primary role should be to kindle and sustain your enthusiasm for the subject. To learn—to really learn—*you* are going to have to do the cerebral heavy lifting. The good news is that—as I hope you've experienced yourself by this point in the book—this heavy lifting can be hugely rewarding.

Derek Muller, the charismatic physicist behind the hugely popular YouTube channel Veritasium, has written and spoken at length about the key role that confusion plays in learning. He points out that when students watch a short video that clearly explains a topic, they really don't learn very much. Deep understanding of a concept requires some degree of mental challenge. This is also why we academics don't provide students with fully worked solutions to every problem we pose. It's much too easy to just look at the solution and say (and even believe), "Oh, that's simple. I could have done that." The real learning happens when you've been struggling with that particular problem for a number of hours without the solution on hand and finally make the conceptual leap necessary to solve it.

What's troublesome is that social media drives us ever further into a clickbait culture where attention spans grow ever shorter; simple, marketable, and meme-able messages are where it's at. This has profound implications for education. To make conceptual leaps we often have to experience frustration and persevere, something entirely at odds with the ethos of instant gratification pervading the virtual world.

Does the world really need another YouTuber who can copy John Petrucci's solos down to the last alternate, swept, or hybrid pick stroke? Or yet another singer who can sound exactly like Rob Halford or Bruce Dickinson? Wouldn't something a little more inventive and original be at least slightly more interesting? When it comes to learning to play guitar, if you can see just how every single note is played, you can mimic what's on-screen down to the last fractional microbend. There's no uncertainty as to just how *that* descending scale or artificial harmonic was played; no confusion. And that confusion, that stumbling around in the dark, can be essential in not only developing an individual style of playing but in developing a deeper understanding of music theory.

There are a number of landmarks in metal history that irrevocably changed the genre. We can list key albums—*Black Sabbath, The Number of the Beast, Master of Puppets, Reign in Blood,* etc.—but there are also many pivotal moments when an individual musician raised the bar for all who followed. They did this almost invariably by finding their own, unique voice: Blackmore's classically inspired noodlings in Deep Purple and Rainbow; Malmsteen's single-handed invention of the neoclassical genre of metal guitar soloing; Steve Vai's otherworldly guitar madness; Philthy Taylor's double bass drum pummeling on *Overkill*[2] and its role in the development of thrash metal via Dave Lombardo and Lars Ulrich; Rob Halford and Bruce Dickinson's quasi-operatic vocal styles.

And, of course, there's the eruption of Eddie Van Halen on *that* platinum-selling eponymous debut album from 1978. Guitar playing would never be the same again, both in and out of metal. Two-handed tapping, dive bombs, legato flurries and all, Van Halen shredded with style, innovation, taste, and *imagination*. Not only did he revolutionize guitar technique, but there was a single-minded pursuit of *his* sound; EVH has been described as a "mad scientist" among guitarists because of his fixation on achieving the tone he wanted. This extended as far as building his own guitars and taking a rather untraditional approach to

---

[2] I remain convinced that Taylor, who was a huge Thin Lizzy fan, was in turn inspired by the double bass drum overkill of Brian Downey on Lizzy's "Sha La La" (released five years earlier).

setting up his amplifiers (including the use of a Variac, a transformer that could reduce the voltage input to the amp from the US standard of 110 V/120 V).

Just a Phase We're Going Through?

A major part of what came to be known as the Van Halen signature "brown sound" (a term EVH coined himself) was due to the contribution of a phaser in the guitar effects chain. It's difficult to describe the sound of a phaser accurately in words—"swooshing," "modulating," "sweeping," and "whooshing" are all reasonable descriptions, but nothing beats listening to the effect itself in action. So dig out *Van Halen* (I) and take a listen to the intro to "Ain't Talkin' 'bout Love" or "Atomic Punk," archetypal examples of phasing in action.[3] Although Van Halen is arguably most associated with the effect, it appears extensively throughout metal and rock music, ranging from its use by the guys in Korn to, at the less heavy end of the spectrum, Dave Gilmour's subtle and accomplished (as ever) harnessing of the sound for Pink Floyd's "Have a Cigar." And phasing isn't restricted to guitar. Jimmy Page and Robert Plant would be appalled to be linked in any way with heavy metal (including being cited in a book with "metal" in the [sub]title),[4] but it has to be said that Zeppelin memorably exploited phasing for John Bonham's massive drum sound on the epic "Kashmir" (listen to the cymbals in particular), while Thin Lizzy's "Dancing in the Moonlight" features a phaser-drenched bassline from the exceptionally talented and sorely missed Phil Lynott.

---

[3] Or, alternatively, let Colin here play you a few examples of phasers in action: https://m.youtube.com/watch?v=vDMVORAJjzE.

[4] . . . to the extent that they refused to appear in Sam Dunn's marvelous *Metal Evolution*—a twelve-part series documenting the origins, history, and growth of metal music. If you've got even a passing interest in the genre, you should watch Dunn's informative and entertaining documentaries. His love of the music, coupled with an encyclopedic knowledge of the various metal subgenres and perceptive analyses, makes for engaging viewing.

So how does a phaser work? How does it process the raw signal from the guitar (or bass, or drums, or vocals) to produce that signature swooshing sound? As the name might suggest, it's got a lot to do with phase. Back in Chapter 3 we looked at the phenomenon of phase in the context of Marsha running around her perfectly circular mosh pit, but since then phase differences have taken a backseat—we had too many other facets of waves to cover. It's high time we gave phase the attention it deserves, because it's right at the core of the widely advertised wackiness of the quantum world.

As you'll remember, the phase difference between two waves is just a measure of how much their peaks and troughs line up (or not). Here's an example to jog your memory:

In the evocative still above (taken from their most recent video), Metallizer's frontman, Vulgus Magna, and his brother-in-arms, Mr. Volumus,[5] are whistling the opening note to their epic new concept piece, a symphonic metal version of the Andrew Lloyd Webber musical *Whistle Down the Wind*. Let's assume that they've both got perfect pitch (in all senses of the term) and they produce exactly the same note. The

----

[5] Less the Toxic Twins, more the Dubious Duo . . .

only difference between their two notes is that Max started to whistle just fractionally before Vulgus. (Guitarists always have to jump the gun.) This means that their notes are slightly out of sync. In this case Max was a quarter of a wave ahead of Vulgus. That's a phase shift of $\pi/2$ radians (or 90 degrees if you prefer the Babylonian unit).

Now, our ears really can't pick out absolute phase differences: if I play you two waves one after another that have a slight phase difference, you'll not perceive any change. What we can detect, however, is the *interference* between two phase-shifted waves when they run into each other (and subsequently encounter our eardrums). As we've seen before, when the peaks of the waves align, there's constructive interference, whereas when the peaks line up with the troughs, we have full destructive interference— the waves cancel each other out at that point. Destructive interference of this type is exploited in noise-canceling headphones, where a phase-shifted noise signal is added to the waveform reaching your ears in order to cancel out the ambient noise (to a greater or lesser extent).

The phaser takes the manipulation of phase to a higher level. Here's how it works:

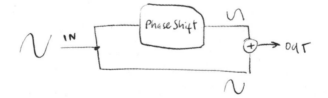

That simple schematic of course doesn't reveal the sophisticated analog and/or digital jiggery-pokery involved in processing the signal, but it nonetheless captures the core idea and the essential aspects of the processing pathways. First, the signal from the guitar is split in two. One of these paths remains uneffected (and unaffected), while the other suffers a phase shift. The clever aspect of the phaser is that the phase shift depends on the frequency of the signal; different frequencies have their phase shifted by different amounts. The signals are then recombined.

The upshot of mixing the raw and phase-shifted signals is that for certain frequencies they'll be in step, while for others, the peaks will be offset from each other. To best see the effects of this type of wave interference on the output of the phaser, we again turn to Fourier and plot a frequency spectrum. Here's how the frequency spectrum (this time frequency is on the y-axis) evolves with time (on the x-axis) for the tapped arpeggio sequence in Eddie Van Halen's "Eruption"[6] (which kicks in at about the 00:55 mark), with and without the phaser effect added:

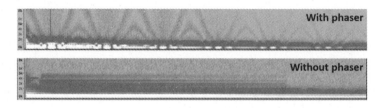

Those notches in the spectrum arise from destructive interference for specific ranges of frequency where the phase difference between the waves is just right for cancellation (or, at least, partial cancellation) to occur. When the waves are in—or out of—step in just the right way and at just the right frequency, they cancel out, producing a notch in the frequency spectrum. It's those notches, and their variation in time, that give rise to the characteristic swoosh of the phaser.

An alternative way of manipulating phase is the flanger, the phaser's partner in crime. Iron Maiden gives us an excellent example. Dial up "The Number of the Beast." First savor that wonderful "Woe to you, oh Earth and Sea . . ." intro and Dickinson's masterful creation of just the right mood via his vocal performance. Then pay particular attention around the 00:44–00:51 mark. That "swooping" sound you hear is a flanger going about its business. The opening swirl of Rush's "The Spirit of Radio," the choruses and outro of Ozzy's "Flying High Again,"[7] the galloping of Heart's "Barracuda" riff, the intro to Rage's "Killing in

---

[6] Well, my poor rendition of it, at least.

[7] Randy Rhoads's addition of a flanger to his effects chain here was particularly appropriate given that the sound is often described in terms of a jet plane flying overhead.

the Name," and the final few seconds of Judas Priest's "Painkiller":[8] all examples of the flanger in action.

While the etymology of "phaser" is fairly straightforward, the origin of the term "flanging" is somewhat less transparent. Like its phasing counterpart, the flanging effect results from the mixing of two signals, but for the latter a uniform time delay, rather than a variable phase shift, is introduced. The effect was first created back in the 1960s (or perhaps earlier) by the sound engineer placing a finger on the rim (flange) of a tape deck, thus physically slowing one version of the recording in relation to another, unperturbed, reel of tape. The origin of the term "flanging" is disputed but it's generally attributed to John Lennon, in response to George Martin's wonderfully tongue-in-cheek explanation of the effect: "Now listen, it's very simple. We take the original image and we split it through a double vibrocated sploshing flange with double negative feedback."[9]

As the Metallizer duo admirably demonstrated previously, a time delay also introduces a phase shift; the flanger and the phaser are distinguished by how the phases are manipulated. I would argue that there's a much more mechanical, artificial, even metallic "feel" to a flanger, as compared to the softer "whooshing" of a phaser, but your ears might perceive things differently. The metallic timbre of the flanger might help explain its greater popularity in metal, however.

Whichever effect we choose, flanger or phaser, we're exploiting the interference that results when there's a phase difference between two waves (or two signals). And constructive-versus-destructive interference is at the very core of quantum mechanics. It's what ultimately, fundamentally, distinguishes the classical physics we experience every day "up here" from the weird, counterintuitive world that exists down there.

---

[8] I know they're not metal, but I also have to give an honorary mention to one of my favorite bands, XTC, for their use of flanged drums at the start of "Making Plans for Nigel."

[9] Wikipedia attributes this quote as being taken from George Martin and William Pearson's *Summer of Love: The Making of Sgt. Pepper*, published by Pan Books back in 1994.

Down at the quantum level, it's all about relationships. Phase relationships. From a particle's perspective, being out of phase with one's surroundings can be a matter of life or death.

It's Always a Mystery . . .

The Dio fans among you will recognize the lyric above. I suspect that Ronnie didn't have quantum physics in mind when he wrote the chorus to "Mystery," but every time I hear that classic track from *The Last in Line* I'm put in mind of this memorable quote:

> *We choose to examine a phenomenon which is impossible, absolutely impossible, to explain in any classical way, and which has in it the heart of quantum mechanics. In reality, it contains the only mystery. We cannot make the mystery go away by "explaining" how it works. We will just tell you how it works. In telling you how it works we will have told you about the basic peculiarities of all quantum mechanics.*[10]

Yes, it's Feynman again. He's talking about the iconic double-slit experiment, once voted the "most beautiful experiment" by the readers of *Physics World* (the flagship magazine of the UK Institute of Physics and, I quote, the "world's leading physics magazine"). We're going to see that the double-slit experiment has deep parallels with the physics of the phaser. The double-slit setup is, in fact, much less complicated than a phaser, save for one exceptionally important complication. When it comes to phasing a guitar riff, we know where the waves come from: the string vibrates, the wave motion is turned into an electrical signal by the pickup, the electrical waves are processed by the phaser, and the signal is then converted from an electrical to an acoustic signal.

For quantum waves, however, the jury is still out as to whether or not they have a physical, objective "reality": debate rages among physicists,

---

[10] From *The Feynman Lectures in Physics Vol. III*, by Richard P. Feynman and Robert B. Leighton, published in 2011 by Basic Books.

philosophers, and mathematicians as to the ontological vs. epistemological nature of the quantum wave function. Is the wave function telling us about what's objectively "out there," or is it rather a measure of what we know (or could know) about nature?[11]

What we *do* know is that with a little mathematical manipulation, we can convert those quantum waves into probabilities. (You may recall the map of Metallizer's ticket sales a few chapters back.) To ensure that our mathematical predictions and analyses agree with our observations and experiments, we have to take interference into account. Getting those probabilities right means considering how the peaks and troughs of our quantum waves align.

Before we dive into the parallels between the double-slit experiment and phasing/flanging, let's first consider just one slit. Actually, no, scrub that. Let's step back even farther and start with no slits at all. Let's consider light just bouncing off a surface. In a suitably metal context, of course . . .

A Mirror Is a Negative Space[12]

Though I certainly appreciate their contributions to the canon of heavy metal, I was never really into Blue Öyster Cult. I don't quite know why.

---

[11] This ontological-vs.-epistemological tug-of-war has characterized quantum physics debates right from the very start. In the Copenhagen interpretation, which remains the preferred flavor of quantum mechanics among physicists (by a small margin, as reported in a *New Scientist* article in 2017: https://www.newscientist.com/article/mg23331074-600-physicists-cant-agree-on-what-the-quantum-world-looks-like/), the question of what happens before a measurement is made is brushed aside. Actually, it's more than brushed aside—the question makes no sense in the worldview of the Copenhagen interpretation. For the "Copenhagen-ists," the measurement is all that matters; we don't know what happens before the measurement so can only speculate, and speculation without supporting evidence is not science. For those who would like to dip their toes into the philosophical battleground that is the nature of the wave function, I thoroughly recommend Matthew Leifer's comprehensive review, "Is the Quantum State Real?", https://arxiv.org/pdf/1409.1570.pdf.

[12] From Blue Öyster Cult's "Mirrors." *Mirrors.* Columbia Records, 1979.

Throughout their lengthy career, they've specialized in brilliantly bonkers song (and album) titles, off-the-wall lyrics, and gargantuan riffs. That should have been more than enough to make me a fan for life. I mean, who could argue with titles like "Veteran of the Psychic Wars," "Fire of Unknown Origin," "I'm on the Lamb But I Ain't No Sheep," and "Cultösaurus Erectus"? *And* they've got an umlaut in their name. It doesn't get much more metal than that.

Leaving aside my lapse in musical appreciation, I've brought BÖC into the narrative at this point because they were one of the first bands to use lasers extensively in their stage shows. (As is common knowledge, all physicists get a warm, fuzzy feeling when in close—or, indeed, not-so-close—proximity to a laser.) Back in the 1970s, Eric Bloom, one of the band's guitarists, pioneered a specially designed fiber-optic bracelet that he used to fire a laser beam at a mirror ball, scattering the light into the audience. As Bloom's partner in crime, "Buck Dharma," discussed in an interview with *Vintage Guitar*, this innovative use of laser technology was not, however, universally appreciated at the time:

> [T]he government flipped out when they realized what was going on; they didn't like the idea that rock and rollers had all of this "power," so they made us invent a lot of fail-safes and interlocks for the equipment. *OSHA* [the US Occupational Safety and Health Administration] actually followed us around on tour for three months! After that, they even clamped down on our "scan" effect, which was a laser cone that was also stunning, but there's no danger as long as the laser is scanning. You see that effect in movies now, but you can't see it at live concerts anymore. Anyone who saw one of our laser shows back then saw something that isn't done at concerts now.[13]

Here's an artist's impression of Mr. Bloom's laser beam extravaganza in action:

---

[13] From the August 1998 issue of *Vintage Guitar* magazine.

We don't need to consider something quite as complicated as a mirror ball, however, to get our heads around the importance of phase in quantum mechanics. A garden-variety mirror is more than good enough. (Indeed, we don't even need a mirror; we could think about reflections just from the glass of a window.[14]) You may remember from secondary/high school science class that when light reflects off a mirror, the angle of incidence is equal to the angle of reflection. This is a core axiom of what's called ray, or geometric, optics, but if you've forgotten about it, don't worry—the illustration on the next page should clear things up. We exploited this principle all the way back in Chapter 2 when we tracked the motion of the bass drum beater, when the light from the laser pointer bounced off the mirror on the beater.

---

[14] The partial reflection of light from glass is a fascinating phenomenon. Feynman teases out the physics in *QED: The Strange Theory of Light and Matter* (Princeton University Press, 1985). If you haven't read it, I'd advise that you put this book down right now and hunt down a copy of Feynman's book. You'll then realize from whence inspiration for this section of the book came.

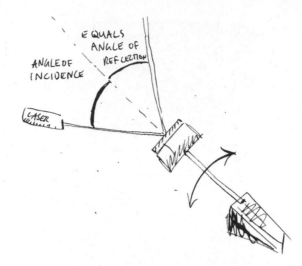

But by now we know that light isn't really just straight-line rays, despite diagrammatic appearances to the contrary. The "atoms" of light are photons: elementary particles that are the quanta of the electromagnetic field. That electromagnetic field is made up of electric and magnetic waves oscillating millions upon millions of times per second. And, like any wave, light can interfere, constructively and destructively; the rainbows we see on CDs and DVDs are just one of the more obvious examples.

We've seen that phasers and flangers exploit phase differences in sound waves,[15] producing characteristic notches in the frequency spectrum. Now, while light is a very different beast from sound in many ways—it doesn't need a medium to travel; it's a transverse, as opposed to a longitudinal, vibration; and its characteristic speed and frequencies are so fast as to leave sound stranded on the starting blocks—at heart they're both waves.[16] This in turn means that the phase manipulation at the heart of the rhythm guitar intro to "The Number of the Beast" (and all those other classic phased/flanger riffs) can be echoed—or should that be mirrored?—optically.

---

[15] Well, more technically, for sound waves that have been converted to electrical waves and back again.

[16] And particles. Yes, even sound acts like a particle. The counterpart of the photon for sound waves is the *phonon*. More on this at the end of this chapter.

You might think that optical phase manipulation requires instruments orders of magnitude more expensive than the humble phaser or flanger that forms part of the metal guitarist's armory. For sophisticated quantum optics experiments that probe the fundamental behavior of single particles, this is certainly the case. But I am confident that you've also seen a cheap-as-chips optical phaser in action many, many times. Those beautiful rainbow patterns that appear in the oil films atop puddles on a rainy day? That's an optical phaser in action. All that's required is a bit of water and a bit of oil. The oil film spreads out so that its thickness is comparable to the wavelength of light: about 350 to 700 nanometers. Imagine we have just a single wavelength of light striking, and reflecting off of, the film—for instance, let's say we're shining the beam from a laser pointer onto the oil. Some of this light will reflect from the top surface of the oil film, while some keeps going and reflects from the bottom (the amount of light reflecting from each can be worked out using what are known as the Fresnel equations—luckily, however, we needn't go into the gory detail of Fresnel analysis in order to understand the physics behind an oil rainbow). Waves that travel through the film before being reflected back out have a longer distance to travel than those that strike the upper surface and turn around there and then. Like this:

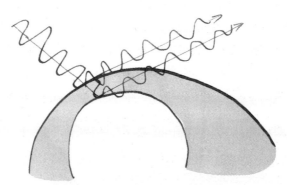

This difference in distance traveled means that the waves reflected from the top surface of the oil will be in, or out of, step with those reflected from the bottom. In other words, there'll be a phase difference, and therefore the two optical paths can constructively or destructively interfere.

But unlike the light from a laser pointer, white light isn't made up of just one wavelength: instead, we have a range of wavelengths. The extent to which the light waves are in, or out of, step now depends on two things: the thickness of the oil film and the wavelength of the light. The film's thickness is just right to cause constructive interference for some wavelengths; for others, the reflected waves will be out of step by the correct amount to cause destructive interference and that color will be "cancelled out." The net result is vivid bands of color on the oil slick: a rainbow in the dark.[17] The white light shining on the puddle is separated into its component colors not by refraction (the bending of light of different colors by different amounts), as in a prism or a raindrop, but by interference effects. It's *precisely* the same physics at work in the phaser (and flanger): certain frequencies are pruned from the spectrum because the phase relationship between the corresponding waves produces destructive interference.

Set Phasers to Stun

Having considered reflections and phase for a wobbly oil film, let's now get back to the case of a perfectly flat reflecting surface—that bog-standard mirror I mentioned above. This time, reflections happen only at the surface itself; there would seem to be no interference effect at play here, unlike the oil film. So does that mean phase differences don't play a role in the reflection process? As a certain New York thrash band would have put it back in their heyday: NOT. Feynman would have been similarly emphatic—in part because he, like Anthrax, hails from Queens, NY. (Feynman's New York accent and mannerisms were often

---

[17] Please excuse yet another Dio allusion. Given Ronnie's penchant for rainbows, however, this chapter could have been packed wall-to-wall with references to the man's work.

described as being at odds with the supposedly more "cultured" or refined speaking styles of some of his colleagues.)

In quantum physics, phase is everything. Feynman demonstrates this exceptionally well in *QED*, where he shows that constructive and destructive interference are at the core of how not only light, but every quantum mechanical particle (or wavicle, if you prefer), behaves. In the case of reflection from a mirror, the classical physics that you learn at school, and that we experience every day—angle of incidence equals angle of reflection—is, in fact, the result of the interference of all possible paths that light could have taken when bouncing off the mirror's surface.

That's right, *all* possible paths. The result we see in the big, bad world around us falls out of the quantum because only those photon paths that interfere in the right way contribute to the final result; destructive interference "kills" the other paths, and we're left with the angle of incidence = angle of reflection rule. Zoom down to the quantum level, however, and we've got to take into account all possible paths and consider how the probabilities stack up.[18]

But I don't want to reinvent the wheel here. There are many great books out there dissecting just how reflection works at the quantum level. Foremost among these is Feynman's own *QED: The Strange Theory of Light and Matter* (as mentioned earlier), but running a close joint second are John Gribbin's *Computing with Quantum Cats: From Colossus to Qubits* and Chad Orzel's marvelous *How to Teach Quantum Physics to Your Dog*. Unlike Orzel, my primary motivation is not canine education, laudable though that is. Our focus is on teasing out correlations between metal and the quantum world, and you'll recall I mentioned that phasing of guitar riffs has a lot in common with the double-slit experiment. How so?

---

[18] This is the essence of Feynman's path-integral approach, and it stems from an after-class conversation he had as a teenager with his high school physics teacher, Mr. Bader, about the principle of least action. Dig out *The Feynman Lectures on Physics* if you want a detailed and thoroughly fascinating explanation.

As we've discussed, when Eddie Van Halen runs the waveform from his guitar pickups through his MXR 90 phaser pedal, the signal is split in two, and it's the interference of these two signal waveforms that provides that essential ingredient of his signature brown sound. Similarly, in the double-slit experiment, light from a single source—the optical analog of the signal from Eddie's guitar—encounters two slits, which provide separate pathways for that light. Just as in the audio phasing that flavors "Eruption," the interference of these signals determines the final result.

The iconic double-slit interference pattern results from the phase differences between the varying paths light follows; as ever, it all depends on just how much the waves are in, or out of, step. If you haven't seen a double-slit interference pattern before, here's what it looks like, alongside a sketch of the setup needed to produce it.

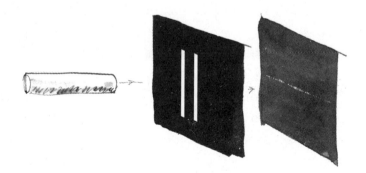

If you've got a laser pointer, some gaffer/duct/insulating tape, and a piece of wire on hand—along with a little patience—you can set up the double-slit experiment at home. The great thing about the easy availability of laser pointers now—as compared to when Thomas Young, "the last man who knew everything,"[19] first did the double-slit experiment back in 1801—is that laser light is *coherent*. Unlike the light

---

[19] See Andrew Robinson's *The Last Man Who Knew Everything: Thomas Young, the Anonymous Genius Who Proved Newton Wrong and Deciphered the Rosetta Stone, among Other Surprising Feats* (Pearson Education, 2007).

from a standard incandescent bulb or a fluorescent tube (or the sun, for that matter), the optical waves output from a laser are all in step with each other. They're entirely in phase, like thousands of moshers racing around a circle pit in perfect synchrony. Other nonlaser sources—candles, torches, bulbs, LED lighting, daylight—are *incoherent*: the phases don't match up and the waves are out of step. Think highly coordinated circle pit vs. the random, uncorrelated motion of a mosh pit.

As with Feynman's mirror example, at the most fundamental quantum mechanical level we'd need to take into account all of the paths that the light follows as it makes its way from the source to the screen on which the final pattern is formed. But just as all of that sophisticated quantum mechanical path integration reduces down to a very simple bit of physics (angle of incidence = angle of reflection), the double-slit interference pattern can be predicted using nothing more than a ruler and some high school trigonometry.

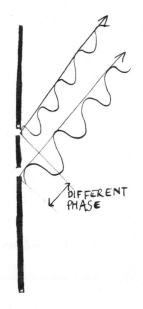

But this type of geometric analysis, while straightforward, can be exceedingly tedious. After all, to get a picture of the whole pattern, we'd need to determine how the waves interfere not just for one point on the screen but for a large number of points. That's irksome enough with two slits. Imagine if you had three, four, ten, or twenty different slits, each with its own phase shift that needs to be taken into account at multiple positions. It becomes very tricky indeed to keep track of all of those different pathways. Sure, you could use a computer to carry out all of those tedious calculations—computers just love tedium—but there's a much more mathematically graceful way to discern the pattern. Guess which nineteenth-century French mathematician and physicist is about to make a (re)appearance?

Yep. Enter Fourier. Yet again.

The diffraction pattern we see on the screen in the double-slit (or, indeed, single-, triple-, or 10^{100}-slit) experiment is the Fourier transform of the slit pattern. It's a breathtakingly elegant piece of mathematical physics. Forget about those tedious geometric calculations. All we need is Fourier.

You've probably not yet managed to shake off the slightly disturbing image of those striped spandex strides that we dissected in an earlier chapter, and you'll remember that we used Fourier analysis to "translate" the stripes into a representation in terms of frequency by working out the spacing of the stripes. We're going to revisit those stripes because, remarkably, they're all we need to understand the double-slit interference pattern (or a single-slit pattern, or an n-slit pattern, where n is any integer we like). Stryper may be responsible for some of the worst crimes against sartorial taste and decency committed in the 1980s—and they were up against some stiff competition—but their fashion choices are an invaluable tool when it comes to explaining quantum physics.

Let's start by thinking about just one stripe. The sketch on the next page shows the outline of what's called the transmission function for a single slit, as might be used to diffract light in an interference experiment. The transmission function describes how much light is, er . . . transmitted. For a single slit it's all or nothing: the transmission function is 0 for where the light is blocked and 100 percent for where the light passes through. I've laid this function over our single stripe, and you'll notice it lines up neatly with the variation across the stripe. We saw the same shape in the function we used to describe the pattern of

Stryper's stylish leggings in Chapter 8. Physicists call this kind of variation a top-hat function, for reasons that I hope are clear.

DO TRY THIS AT HOME

It's even easier to do a single-slit, as compared to a double-slit, experiment at home. Grab the laser pointer from the double-slit experiment you set up a few pages back and then find a piece of cardboard. Cut a square hole in it, a few centimeters on each side. Then find a long strand of hair. I'm suspecting that this won't be an issue for many reading a book about metal, but for those who, like myself, may be more follicularly challenged now than when they first dropped the needle onto *Screaming for Vengeance*, *Strong Arm of the Law*, or *Piece of Mind*, don't worry. A thin piece of wire will do the job just as well.

Place the strand of hair (or wire) across the hole in the card, make sure it's nice and taut, and then tape the ends down so that it's held firmly in place. Find a wall that you can use as an "observation screen," and then shine the light from the laser pointer through the hole onto the hair. You'll get a single-slit diffraction pattern on the wall. (For reasons I won't go into, the diffraction pattern for a thin solid object like a hair is identical to the pattern you get for a slit of the same dimensions. Let's just say that it's yet again all to do with phase.)

Although this is too often glossed over in pop-science descriptions of quantum physics, light shining through a single slit also forms an interference pattern. You might, very reasonably, ask why this is. If there's only one slit, where is the interference happening? The wave interference in this case can be understood in terms of phase differences between waves originating from the different ends of the slit. It's a subtly different type of pattern compared to that observed for two slits, but it's an interference pattern nonetheless. We could analyze the paths of light that emerge from the single slit in the same geometric style we used for the double slit, but we don't need to do that. Instead, we just take a Fourier transform of the pattern of the slit and, lo and behold, we get the diffraction pattern.[20] (This is the basis of a huge field of physics known as Fourier optics.) When you first encounter this link between diffraction and Fourier analysis as a physics student, it can seem like witchcraft. The slit does the complicated mathematical calculations for you and effortlessly produces the result without anyone having to spend hours working out the integrals correctly.

Remember what Fourier analysis is all about: it's nothing more than combining waves in the right way to produce a particular pattern. This is what's happening with the slit; it's doing the Fourier "de-synthesis" for us. The diffraction pattern we see on the wall tells us which spatial frequencies are required in order to produce the slit's top-hat function.

---

[20] I can feel a great disturbance in the Physics Force as my expert colleagues get a little twitchy about my description here. Yes, I know there are a couple of additional minor mathematical steps required to get from the Fourier transform (FT) to the diffraction pattern, but the information we need is nonetheless "encoded" in the FT. (And, again, for the experts, I'm of course talking about Fraunhofer diffraction rather than Fresnel diffraction throughout.) As ever, it's the bigger picture rather than the mathematical minutiae I'm concerned with.

Let's compare. Here's the Fourier transform of a single slit—or, equivalently, a single stripe of spandex:

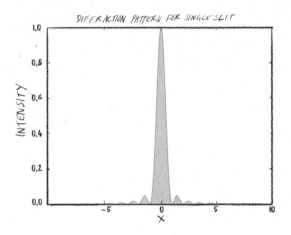

This is *exactly the same* pattern we see in the light intensity measured across a diffraction pattern from a slit of the same width.

And if we narrow the slit width, guess what happens to the diffraction pattern? It spreads out—the central peak in the pattern gets wider (as do all of the other peaks) and the peak separation increases.

It's the same old uncertainty principle, just in a different context: narrow in time, wide in frequency; narrow in space, wide in reciprocal space. The narrower we make the slit, the more the diffraction pattern broadens.

If we go to two slits, or, equivalently, two spandex stripes, this is what we get:

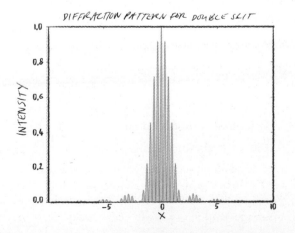

That's the double-slit interference pattern. Note that no complex quantum mechanics is required. We will generate exactly—and I mean *exactly*—the same pattern by Fourier analyzing a couple of stripes on spandex as by shining light from a laser pointer through suitably arranged bits of tape, cardboard, and wire.

We could continue adding slits/stripes and Fourier transforming to get the diffraction pattern each time, but I think you've got the core idea at this point. Let's turn to the really perplexing aspect of the double-slit experiment; the mystery to which Feynman alluded. We've interpreted everything thus far in terms of waves—the interference patterns we see ultimately arise from just how the light waves from the slits mix and match. But there's still that thorny wave-particle duality to deal with.

As we've covered in earlier chapters, light can also be thought of as a stream of photons. We can turn the intensity down until we have a steady drip-drip of photons, rather than the tsunami of light that a laser pointer produces. Indeed, the light intensity can be reduced to the point where photons pass through the slits one at a time. This type of single photon-counting experiment is now routine in physics labs across the world (but, unfortunately, you're not going to be able to re-create this one at home using bits of tape, cardboard, hair, and/or wire).

You might well ask just what happens to the interference pattern when we set up the double-slit experiment so that photons pass one at a time. And you'd not be alone. That deceptively simple question continues to confound physicists. Channeling Feynman yet again, I can tell you what happens in the one-photon-at-a-time version of the double-slit experiment, but I can't explain precisely *why* it happens. No one can. Yet.

If we turn down the light[21] to the point where only one photon is incident on the slits every second (or every minute, or every hour . . .), at first what we see on the screen is a jumbled mess of pinpricks of light. (Because we're using a single-photon counter, we can register the arrival of each photon as an individual spot.) As the photons trickle through,

---

[21] . . . or, as Sabbath would have it, turn up the night.

the number of points of lights on the screen increases accordingly. At a rate of arrival of one photon per second, for a long time it looks as if the distribution of those pinpricks is entirely random, rather like those aerial shots from '80s arena gigs when the multiplatinum-selling band onstage has launched into their latest power ballad and hundreds of lighters are held aloft:

Patiently we wait, all the while capturing each tiny flash of light from each arriving photon. And then, excruciatingly slowly but no less remarkably for that, we see that a pattern is forming out of the haze:

The double-slit diffraction pattern emerges from the noise.

So what? We knew that was going to happen—all you've done is slow down the experiment. It's like running a video at half or quarter speed; it's not really surprising that the ending is the same as when we play the movie at normal speed.

True, but our experiment isn't a movie, and this reasoning overlooks a key detail. What's mind-blowing here is not that we can slow down the rate at which the interference pattern is built up. It's that we can slow down the experiment to the point where there is only ever one photon—a single, lonely photon—passing through the slits. In principle, we could slow the experiment down to the point where one photon passes every year, or every millennium. A single photon, not interfering with or otherwise being bothered by any other. *And yet the interference pattern ultimately still appears.*

This is, to put it mildly, deeply perplexing. We know that the pattern itself is fundamentally due to the properties of waves: the bright and dark patches arise from the way light waves interfere. You are very aware by now that Fourier analysis is the algebra of waves—I'll stress again that the Fourier transform of those two stripes of spandex is identical in every way to the interference pattern formed in the double-slit experiment. An explanation of the double-slit interference pattern simply must incorporate waves. Somehow, even when we ensure that we only ever have one particle of light in the experiment at a time, we still have wave interference.

This is why Feynman said that the double-slit experiment contained all of the mystery of quantum physics.

After many decades of collectively reflecting on this result, most physicists rationalize the one-photon-at-a-time interference pattern in terms of the multiple-path approach that we discussed in the context of mirrors. In other words, we accept that the photon is somehow taking all paths at once; that it's going through the two slits at the same time. Yes, I know that sounds crazy. But if we make this assumption in our mathematical models, they predict what we see in the real world with amazing accuracy. The interference pattern arises from the phase differences between the different paths a photon can take. In other words, the interference can be thought of as arising from single photons passing through both slits simultaneously. Moreover, it's not just photons that do this. The one-particle-at-a-time version of the double-slit experiment has been carried out for electrons[22] and, most recently, molecules.[23] And quite large molecules at that.

For me, there is no better demonstration of the bizarre wave-particle duality of matter than the C_{60} molecule. We work with this molecule a great deal in our research group at the University of Nottingham for a variety of reasons but not least because it's quite literally a nanoscale football.[24] The sixty carbon atoms composing the molecule (hence C_{60}) are arranged in exactly the pattern of hexagons and pentagons found on a conventionally stitched ball, the atoms at the vertices of twenty hexagons and twelve pentagons (as in the sketch on the next page) interlocking to form the most beautifully symmetric molecule in nature.

Being roughly one nanometer in diameter, C_{60} (also known as buckminsterfullerene or the buckyball) is custom-built for nanoscience. Our team has spent a great deal of time investigating its properties like proper, serious scientists. Some of our most enjoyable experiments, however, have involved playing "football" with C_{60} by pushing it around with the tip of a probe microscope. That's shown on the upper right on the next page. You'll see that in the scanning probe microscope images, the buckyballs look exactly like, well, balls—nanoballs that are so solid

---

[22] Search YouTube for "single-electron double-slit experiment." You'll find footage of the stunning experiment carried out by Akira Tonomura and his colleagues at the Hitachi research labs back in the 1980s.

[23] Juffman, Thomas et al. "Real-time Single-Molecule Imaging of Quantum Interference." *Nature Nanotechnology* 7 (2012): 297–300. This is known in the trade as a heroic experiment: exceptionally challenging, requiring painstaking attention to detail, but where the gain is definitely worth the pain. Juffman and his colleagues captured single-molecule-at-a-time double-slit interference patterns for molecules as large as 1,298 atomic mass units. That's a big chunk of matter in the quantum world. And yet precisely the same double-slit interference pattern built up.

[24] For those readers on the other side of the "pond," I should stress that the football in question is a spherical object, not a prolate spheroid.

and well defined that we can push them on a surface, much like we can roll a billiard ball across smooth green felt. (I should note that we have compelling evidence that buckyballs do indeed actually roll when we push them.) Yet also included in the image below is the graphed result of an interference experiment using a beam of C_{60} molecules[25] where, once again, we see the same pattern of fringes as we saw for the double-slit experiment with light. In that case, those buckyballs are behaving just like waves. In other words, a molecule that is about as close as it gets to a billiard ball–like particle in our scanning probe experiments—to the extent where we can even "pot" the buckyball in holes (i.e., missing atoms) in the surface—behaves like a diffuse and ethereal wave when it's investigated in a different way.

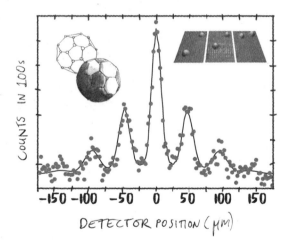

Back in Phase

Ultimately, wave interference is the source of all the weirdness of the quantum world. For that weirdness to manifest itself, to get a clear look at the behavior of individual particles or wavicles and the ways they

---

[25] Nairz, O., Arndt, M., and Zeilinger, A. "Quantum Interference Experiments with Large Molecules." *American Journal of Physics* 71 (2003): 319.

interact with one another, physicists have to work very hard indeed to control experimental conditions and ensure that the phase relationships aren't scrambled. As objects get bigger and bigger—as we move from the atomic to the nanoscopic to the microscopic to the macroscopic—there are more and more ways for quantum waves to get out of step with each other and for the environment to interfere with their interference, so to speak. Or, in slightly more technical language, the waves decohere: that quantum circle pit where everything is neatly in step descends into a much more disorganized moshing mess. We're too damn big, too damn disorderly, and too damn incoherent for the "quantumness" to scale up.

What's rather intriguing is that, despite the obvious importance of phase, scientists often neglect it entirely when breaking down waveforms and signals into their component parts. I'm certainly guilty of this. We've spent the vast majority of this book considering only the frequencies and relative amplitudes of the waves making up signals as diverse as the opening of Metallica's "Sanitarium," Van Halen's "Eruption," top-hat functions of various stripes, and, on the grandest scale, fluctuations in the cosmic microwave background radiation. And not once did a phase spectrum make an appearance.

That may seem strange given that phase is all-important at the quantum level, but in fact there are many situations in science where an analysis of phase can comfortably take a backseat. Often, what's most important is knowing the relative strengths of the Fourier components, rather than the extent to which they're in step. This is particularly true when those components are entirely incoherent to start with—then there's no phase relationship to worry about.

As we've seen, the swooping and swooshing sounds of the phaser and flanger arise not from the variations in phase themselves, but from the effects that those phase differences have on the amplitude of waves of different frequencies. Our ear is not particularly sensitive to phase variations in and of themselves—it's remarkable how much we can scramble the phases of a waveform's Fourier components, completely changing the shape of the signal, and still find that our ears fail to detect a difference.

While we'll soon see how glaringly obvious differences in phase are to our eyes, for our ears all that counts is the mix of the spectral ingredients in the audio signal. Our ears couldn't care less about the order in which we add those ingredients, or how we blend them, or how they're folded into the mix. The sound we hear depends only on the amount of each ingredient. Nothing else matters.

Now, let's close this chapter with a dramatic—some might even say disturbing—example of the key importance of phase differences in how our eyes perceive an image.

Here's a photo (on the left) of Motörhead in their *Ace of Spades*–era prime:

And here's a photo of, brace yourselves, One Direction.

→

I've deliberately selected two photos with very distinct differences in image, style, and culture, for reasons that will become clear very soon.

You'll remember that we can break any image down into its component spatial frequencies. (How could you forget the decomposition of Stryper's spandex glory?) We've focused thus far on amplitude, the strengths of the different Fourier components making up each image—just how much of each spatial frequency is present? But we've entirely neglected to consider how those different spatial waves relate to one another. How do the peaks and troughs line up? In other words, forget the frequency spectrum . . . what does the phase spectrum look like?

I don't need to embark on a lengthy discussion of how we come up with the phase spectrum for a signal (be it a photo of Motörhead, the entirety of *Reign in Blood,* or a map of the CMB).[26] Fourier's mathematics provides the phase spectrum for free—it falls out of the analysis with no extra effort. Let's leave aside just what the phase spectrum looks like for each image because it's rather complex in both cases. Instead, we'll use a time-honored method of demonstrating the importance of phase in image perception. Let's mix the amplitudes of the frequency components for Motörhead with the phases for the One Direction image, and vice versa. Here's what happens:

One Direction amplitudes with *Motörhead* phases

Motörhead amplitudes with *One Direction* phases

That's right, it's not the amplitudes of the frequency components in the images that really matter. It's the phase relationships that count. While our ears are generally indifferent to phase differences—unless they affect the amplitudes of the frequency components via interference—our eyes are very sensitive indeed to variation in phase. So much so that we can mistake Motörhead for One Direction if we get our phases mixed up.

———————

[26] There's a brief explanation in the "Maths of Metal" appendix.

Despite his hard-drinking, hard-living, born-to-raise-hell persona, Lemmy had a steely determination and a focus that were too often over-looked. While he may have often played up the shambolic rock-and-roller aspects, it requires a substantial level of coherence and dedication to take a band like Motörhead to the top (or, indeed, over the top). Similarly, while the audience at many metal gigs may often seem to be chaotic and lacking in any type of underlying organizing principles, appearances can be deceptive. Especially in science, it always pays to look a little deeper into the dynamics.

Chapter 11

CAUGHT IN A MOSH

You aim for someone's head, to stain the floor red
Give someone a kick, to prove you're truly sick
. . . There are some that try, but they won't survive
They don't hit! 'Cos they're wimps!
. . . Everyone's doin' the toxic waltz

—from Exodus's "Toxic Waltz"[1]

To me, [it] takes away from a musical performance when
there's accomplished, talented musicians onstage performing songs for you;
to engage in this totally unrelated physical activity to me is disrespectful,
in a way, to the rest of the audience that doesn't want to do that,
and to the band onstage that are trying to perform for you.

—J. D. Cronise of The Sword, speaking to *Decibel* magazine in 2012[2]

---

[1] From *Fabulous Disaster*. Lyrics by T. Hunting, G. Holt, R. Hunolt, S. Souza, and R. McKillip. © 1989 Music for Nations/Colgems-EMI Music.

[2] See Robert Pasbani's November 2012 post for Metal Injection: www.metalinjection.net/latest-news/drama/the-sword-does-not-want-you-moshing-at-their-shows. See also Fury, Jeanne. "Please Don't Mosh to the Sword, You Disrespectful Sack of Crap." *Decibel* 99, November 27, 2012. www.decibelmagazine.com/2012/11/27/please-don-t-mosh-to-the-sword-you-disrespectful-sack-of-crap/.

over the decades since the subculture decoupled from its rock roots, metal fans have had a reputation in some quarters for being unwashed and uncouth at best, and downright violent at worst. (As if that hoary old "cerebrally challenged" stereotype wasn't enough to endure.) Yet I've always felt a great deal safer at a metal gig than I do strolling through a city center late on a Friday or Saturday evening, and have not once experienced antisocial or threatening behavior directed at me personally in the thirty-three years (gulp) I've been attending metal gigs. That includes sold-out concerts by many of the biggest names in thrash—Slayer, Metallica, Anthrax, Megadeth, Testament, Exodus, Machine Head et al.—who lay claim to having some of the metal genre's more "boisterous" fans.

To those who are not *au fait* with metal traditions, I can understand entirely just why the primal "SLAYYYYERRRRRRR" that erupts from a thousands-strong crowd as four rather imposing gentlemen enter stage left (or stage right) could be somewhat perturbing. But to many of those fans in the crowd, there's something almost comforting about that roar of anticipation. Like so many other phenomena in this book, that roar—or, more accurately, roooarrrrggghhh—resonates deeply. They're home. This is where they belong.

There's a remarkably strong sense of community among metalheads: the denim and leather, the sewn-on patches, the bullet belts, the tattoos, the faded decades-old tour shirts, and the never-to-be-admitted-to spandex strides (striped or not) tucked away at the back of the wardrobe collectively contribute to that almost tribal feeling of belonging. If you're a metalhead, even one, like me, who hasn't donned the trappings of metal in many a year,[3] you can probably identify.[4] That spine-tingling moment when a crowd of ninety-thousand at Donington Park[5]—drenched to the bone and standing knee-deep in torrents of mud because it's been pissing down rain all day—belt out the "Woah-oh-oh-oh" refrain in "The Trooper." Together. A metal communion. (And the allusion to a religious event is entirely deliberate.)

One key aspect of the dynamics of a modern metal gig, however, can be considerably less convivial for some. Metal fans are increasingly divided on the issue of the rather more aggressive forms of crowd activity that have evolved over the years, which originated in the hardcore and punk scenes of the 1980s, and then, via thrash, crossbred with metal. I'm talking about moshing, circle pits, and slam dancing. If you're not familiar with these forms of crowd dynamics at metal gigs, let's just say that they involve a great deal of physicality—running around in circles, slamming into people, and, at worst, injury.

---

[3] Well, other than the T-shirts.

[4] Which is why it was so disappointing that the erstwhile frontman of Pantera dragged metal into the gutter by bellowing "White Power" while raising a Nazi salute at a concert in early 2016. Many in the metal community were appalled by Phil Anselmo's actions (including Robb Flynn of Machine Head who made a lengthy video tackling Anselmo's racist behavior). Dom Lawson, one of the finest writers in metal, wrote an eloquent and memorable riposte in the *Guardian*: "Racism is Phil Anselmo's problem, not metal's." The subheading says it all: "The ex-Pantera singer's antics prove metal has its share of racist idiots, but no more than other genres—and its real purpose is to celebrate the downtrodden, not oppress them."

[5] It's Donington. Donington, dammit. Not Download. That's not to say that the world's foremost metal festival—which went through a rebranding exercise in the 1990s, changing its name from Monsters of Rock to the modish-in-'94 Download moniker—isn't still an amazing event. Donington 2016 featured Sabbath headlining on the Saturday with Maiden as headliner on the Sunday— an awesome festival in every sense of the word. But we old-school metal fans with long memories have a natural aversion to the "Download" brand. Or at least this one does.

Proponents of moshing and slamming argue that it's all just good old "friendly violent fun" and either that no one really gets badly hurt or, if they do, it's their own fault for being in the pit in the first place. Those lyrics to Exodus's "Toxic Waltz" that open this chapter, however, are rather too close to the truth of the matter for the more aggressive fans in the pit. A small minority may be focused less on music and communion than on releasing their aggression in a very far from friendly manner.

MOSHING DECRIED

A number of bands in the metal, punk, industrial, and hardcore scenes have spoken out about the moshing/slamming culture over the years. Early in the rise of the more aggressive forms of crowd interaction, Fugazi would (very politely) ask those engaging in slam dancing or similar behavior to stop. In some cases, they even dragged the ringleader(s) onstage and asked them to apologize to the crowd. It was made very clear that slamming and moshing were not welcome at their gigs. Consolidated and At the Drive-In likewise took very strong anti-moshing stances.

But many other bands actively encouraged mosh pits and slamming; these more extreme forms of crowd participation quickly became the norm. And then, in 1996, moshing tragically resulted in the death of a fan. Bernadette O'Brien was just seventeen years old when she passed away after she was crushed in a mosh pit at a Smashing Pumpkins concert at the Point Theatre in Dublin. A year earlier, the Pumpkins' frontman, Billy Corgan, had made his intense dislike of moshing very clear during a concert in Chicago:

I just want to say one thing to you, you young, college lug head-types. I've been watchin' people like you sluggin' around other people for seven years. And you know what? It's the same shit. I wish you'd understand that in an environment like this, and in a setting like this, it's fairly inappropriate and unfair to the rest of the people around you. I, and we, publicly take a stand against moshing.[6]

Smashing Pumpkins' music is not particularly aggressive, certainly as compared to the heavier, thrashier, grindier end of the metal spectrum. A number of musicians from those types of bands, however, have also spoken out against moshing. Nearly twenty years after Corgan's impassioned admonishments, Slipknot's Chris Fehn was similarly scathing about the levels of violence in mosh pits:

I think, especially in America, moshing has turned into a form of bullying. The big guy stands in the middle and just trucks any small kid that comes near him. They don't mosh properly anymore. It sucks because that's not what it's about. Those guys need to be kicked out. A proper mosh pit is a great way to be as a group and dance, and just do your thing.[7]

Wherever you stand, or fall, on the moshing issue, it's indisputable that metal gigs involve crowd dynamics that are rather far from the norm—mosh and circle pits don't tend to form spontaneously on street corners. From a scientific viewpoint, these novel forms of person-to-person interaction are massively intriguing. Crowd dynamics at metal gigs are complicated, multiparameter, collective, far-from-equilibrium, self-organized phenomena. That collection of adjectives screams, "Here be fascinating physics!" but, remarkably, the statistical mechanics of moshers have only very recently been studied. Initially, you might think that because a mosh pit is such an exceptionally complicated system involving complex human motion coupled with, and constrained by, social dynamics (including the psychology of the various "actors"), it would be nearly impossible to model or simulate that hive of activity. In

---

[6] http://www.mtv.com/news/1434230/fan-crushed-at-smashing-pumpkins-show/.

[7] http://loudwire.com/slipknot-chris-fehn-bullying-mosh-pits/.

many ways, those are fair assumptions. But when it comes to moshing, it turns out that the devil is most definitely *not* in the detail.

Statistical Mechanics of Moshing

For my money, the best title for a scientific paper in the entire history of science is this: "Collective Motion of Humans in Mosh and Circle Pits at Heavy Metal Concerts." Published in May 2013 in *Physical Review Letters*—a prestigious journal in physics circles and beyond—the abstract of the article reads like this:

> *Human collective behavior can vary from calm to panicked depending on social context. Using videos publicly available online, we study the highly energized collective motion of attendees at heavy metal concerts. We find these extreme social gatherings generate similarly extreme behaviors: a disordered gaslike state called a mosh pit and an ordered vortexlike state called a circle pit. Both phenomena are reproduced in flocking simulations demonstrating that human collective behavior is consistent with the predictions of simplified models.*

I was so impressed by this study of the metal-physics interface that I contacted Jesse Silverberg, the first author and lead researcher. Jesse now works at the Wyss Institute for Biologically Inspired Engineering at Harvard, where—as in his work on the physics of moshers (which was done at Cornell)—his research spans a range of exciting topics that bridge disciplinary boundaries; it's difficult to know when the physics and chemistry stop, and the maths and biology begin. Jesse and I exchanged a number of emails and I was struck by the careful, systematic, and insightful quality of the analyses of metal crowd dynamics he and his coworkers had undertaken.

One of the many fascinating aspects of the "Collective Motion of Humans in Mosh and Circle Pits . . ." paper is that it compellingly demonstrates that moshers at metal gigs can be treated just like—and I don't quite know how to break this to you metal fans—dumb, directionless molecules in a gas. The physics of the crowd behavior stemming from moshing requires no consideration of human sociology, interactions, or intelligence. Rather, moshers behave exactly as if they were billiard ball particles—"hard spheres" in the language of a physicist or chemist—simply bouncing into, and rebounding off, each other, scrambling their velocities with each collision.

How do Jesse and his team know that they can describe a mustering of moshers[8] as a "disordered gaslike state"? Well, they did the maths. And, of course, the measurements.

To understand what Jesse et al.'s painstaking analysis of video footage of the crowds at metal gigs revealed about mosh dynamics, I'm going to have to introduce a smidgen of thermodynamics. But don't worry. So far, via the medium of metal, you've already gotten your head around Fourier analysis, quantum physics principles, delta functions, reciprocal space, and the evolution of the universe as imprinted on the cosmic microwave background. A small helping of thermodynamics is therefore not going to *phase* you at all.

Engendered All That Being Hath[9]

Thermodynamics is all about the relationship and interplay of different forms of energy and the methods we can use to measure and "mine" that energy. Given the close connections between matter and energy— and information, but that's an entirely different story (compellingly

---

[8] I couldn't decide on which collective noun to choose here. A mob of moshers? Naah, too pejorative. A mischief of moshers? Doesn't quite cut it. A murmuration of moshers? I like the alliteration, but it doesn't quite accurately describe the randomness of the moshing. (I'll keep it in mind for later on in this chapter, however.) A mustering of moshers seems about the most apt I can find. Feel free, however, to send in your own suggestions on the back of a stamped, self-addressed envelope, as they used to say when I was a denim-and-patch-wearing lad. Or email will do.

[9] From "Molecular Evolution" by the Scottish scientist and poet James Clerk Maxwell (1831–1879).

covered in James Gleick's *The Information*[10])—thermodynamics is absolutely core to our understanding of the universe. The subject had a baptism of fire in the mid-eighteenth century as combustion engines and their steam-driven counterparts fueled the Industrial Revolution. In the early nineteenth century, the French engineer and physicist Sadi Carnot, whom many consider to be the father of thermodynamics, not only laid the groundwork but established much of the framework of the entire field (extending right up to the present day), with his *Reflections on the Motive Power of Fire*.[11]

Despite the strong conviction of many scientists (and nonscientists) that innovations in technology begin with understanding the underlying science, thermodynamics really is the exception that proves the rule. Steam engines, whose development was powered, in all senses of the word, by the insights and dedication of James Watt (in turn extending and improving upon the atmospheric engine invented by Thomas Newcomen at the start of the eighteenth century), predated the science of thermodynamics. The engines were invented, and only *then* were their fundamental operating principles explained by science, not the other way around.

This runs counter to the narrative scientists generally like to reel out, often when it comes time to applying for funding, that scientific insight drives all technological development. Sure, sometimes it works

------

[10] Gleick's tome on the genesis of our current information age was first published in 2011 by Pantheon Books.

[11] While we're on the subject of the motive power of fire, I need to take a brief time-out to flag a major scientific inaccuracy, related to the prehistory of thermodynamics, which appears on a classic metal album and has long irked some of the more pedantic metal fans out there. The record in question is none other than Iron Maiden's *Piece of Mind*, released in 1983. Nestled around the middle of side two—I wore out the vinyl as a teenager—we find a lesser-known Maiden ditty that goes by the name "Quest for Fire," which was penned, like so many others, by the band's bassist, Steve Harris. Here's the beginning: *"In a time when dinosaurs walked the earth, when the land was swamp and caves were home. In an age when prize possession was fire, to search for landscapes men would roam."* Let's just say that Mr. Harris doesn't quite get the chronology correct here.

like that. But, often, a new device is invented or discovery is made and the role of science is to "reverse engineer" the process: to work out the scientific principles that underpin the new piece of technology.

The preferred modus operandi of physicists is to break a system down into its smallest parts: to reduce a complex problem into core "components" and then work out how those components interact to produce the observed behavior. That's why so many physicists, including yours truly, are most at home when considering single particles, atoms, and/or molecules. That reductionism has a deep appeal; it's one of the reasons that those single-atom manipulation experiments I've described in earlier chapters are so much fun. It's also why particle accelerators have expanded in size so much over the years—the bigger they get, the higher the achievable energy, and the more smashing and smushing of fundamental particles they can do. Physicists break matter down into smaller and smaller chunks, isolate them, and study them. When it comes to steam engines, alas, the methodology that has so often served physics well runs out of steam.

A reductionist approach to thermodynamics, whereby individual particles are tracked and classified, would fail catastrophically. Taking Watt's engine as an example, steam, like any gas, is made up of gazillions of molecules darting hither and thither at hundreds of miles per hour. The air around us is likewise thronged with molecules of nitrogen, oxygen, carbon dioxide, and the other constituents of the atmosphere, dashing around and slamming into everything they encounter. If we could follow the motion of even a tiny fraction of those molecules, it would make the most chaotic of metal gigs look positively tame in comparison.

In thermodynamics, substances are measured via a unit known as the *mole*. Much as I'd like to tell you that the etymology of the word has something to do with furry subterranean creatures, its origins are a little more prosaic. The term "mole" is an anglicization of "mol," first coined by the German chemist Wilhelm Ostwald, and an abbreviation of the German word *Molekül*. A mole of gas—or any substance—is the quantity containing 6.022×10^{23} molecules. That staggeringly large number is known as Avogadro's constant, named in honor of the Italian chemist (and erstwhile lawyer) Lorenzo Romano Amedeo Carlo Avogadro (1776–1856).

You might quite reasonably ask just why Avogadro's constant has that particular, seemingly arbitrary, value. In fact, it's not arbitrary at all (though it has been redefined a few times over the years, leading to some confusion). At the time of this writing, Avogadro's constant, also known as N_A, is defined as the number of constituent atoms or molecules in one mole of substance. More specifically, it is the number of atoms in twelve grams of carbon-12, a particular isotope of carbon with an atomic mass of twelve (made of six protons, six neutrons, and six—nearly weightless—electrons). This concept sets up a relationship between grams and atomic and molecular weight. In other words, Avogadro's constant acts as a bridge between our big, bad, macroscopic world and the nanoscopic realm inhabited by atoms and molecules.

Avogadro didn't work out the value of his eponymous constant, however. Surprising, I know, but science is littered with examples like these. A scientific law or constant is often named after someone who may have been only at the periphery of the work. This isn't quite true for Avogadro, given that he made groundbreaking contributions to our understanding of the atomic and molecular nature of matter (including Avogadro's Law, the forerunner to what's known as the ideal gas law—a cornerstone of thermodynamics), but the numerical value of the constant that bears his name was not determined by the man himself. That honor goes instead to an Austrian high school teacher, Josef Loschmidt.

In 1865, Loschmidt used the developing field of kinetic molecular theory to estimate the number of molecules in a cubic centimeter of gas at atmospheric pressure and room temperature, allowing him to connect the atomic and macroscopic worlds for the first time (a molecule is *tiny* compared to the volume of a cubic centimeter), and it was via his insight that the Avogadro constant was able to be defined.

"Kinetic molecular theory" is just a grand name for a way of explaining the properties of gases by considering them collections of molecules that behave like tiny billiard balls, continually bouncing into each other and into the walls of their surrounding container. In reality, as we've discussed before, atoms and molecules are rather more gregarious than this—they like to interact and form bonds. Or at the very least, when they get close together (but not *too* close), they feel a strong attraction.[12] It turns out, however, that this simplified, idealized model of a gas as a collection of tiny billiard balls that bounce off each other (without any "stickiness" at all) is good enough to capture many principles of thermodynamics. We call this spherical cow version of molecular dynamics "the ideal gas," and it's the cornerstone of much of our understanding of how different forms of energy, particularly heat and work (i.e., the energy required to move, lift, push, or pull stuff using pistons and pulleys), are interrelated.

The ability to quantify the molecules in a given volume and theorize about their collective behavior in general terms is one thing, but working out the behavior of those gazillions upon gazillions of frantic particles is quite another. It was Ludwig Boltzmann, a close friend of Loschmidt's, who realized that the way to tackle the problem of the astronomically large numbers[13] of atoms and/or molecules in microscopic, let alone macroscopic, quantities of "stuff" was to harness the power of statistics. Boltzmann, an Austrian physicist, philosopher, and

---

[12] Feynman—yes, him again—once mused over the following scenario. Imagine there's a pending cataclysm and all knowledge is about to be wiped out. (Let's say the Trump administration finally decided to go full *Fahrenheit 451* and rid the world of all that elite, academic, mainstream, FAKE science that LIES to the people.) What one sentence would you leave to future generations that would contain the most scientific information? Feynman argued that it should be this: "All things are made of atoms—little particles that move around in perpetual motion, attracting each other when they are a little distance apart, but repelling upon being squeezed into one another." That attraction happens even for neutral atoms. It's called a van der Waals force, and it's due to the quantum fluctuations of the electrons of each of the atoms: small imbalances result in tiny electric dipoles on different atoms that attract each other.

[13] "Astronomically large" is something of an understatement. There are more atoms in your body than there are stars in the observable universe.

champion of the atomistic model of matter, pioneered the field of statistical mechanics: he believed in the reality of atoms, realized just how many of them were dancing around "down there," and understood that attempting to account for them all on an individual basis was an impossible task for even the smallest fraction of a mole of a gas.

BRING FORTH WHAT IS TRUE

Ludwig Boltzmann's confidence in the atomic nature of matter was not at all shared by other physicists of the time. At the start of the twentieth century, many physicists simply didn't accept that atoms were, to use the language of the early twenty-first century, "a thing." Boltzmann fought long and hard for his atomistic model and its associated statistics, devoting a great deal of time to not only the underlying physics and mathematics but to the philosophical ramifications, in the face of substantial intellectual resistance from his colleagues and contemporaries. This quote from the man himself neatly sums up Boltzmann's personal philosophy: "Bring forth what is true. Write it so it's clear. Defend it to your last breath."[14]

And Boltzmann indeed defended his beloved statistical mechanics until his last, tragic, breath. In 1905, he committed suicide by hanging himself (while on holiday with his wife and daughter in Trieste, Italy). What makes this story even more poignant is that only twenty years later the atomistic model of matter was universally accepted (or as close to universal acceptance as makes no odds); Boltzmann was entirely vindicated. The imaging and manipulation of single atoms we've discussed in previous chapters, which are now almost routine in nanoscience, are arguably the most visually compelling testament to Boltzmann's genius. They "bring forth what is true" in a way he could never have envisaged: Technicolor maps of atoms distributed in natural and artificial ensembles. It's a shame that Boltzmann hadn't been born just a few decades later so

---

[14] Epigraph at the start of Boltzmann's *Principles of Mechanics*, from the Vienna Circle Collection published by Springer.

that he could have seen with his own eyes the stunning evidence for those atoms he so confidently predicted were there all along.[15]

Boltzmann spent a great deal of his life considering how atoms and molecules are distributed within substances, both in terms of their positions and their speeds.[16] As with any system or process involving a large number of entities, whatever they may be, a statistical approach is key to understanding. Boltzmann deserves the lion's share of the credit for injecting this statistical mind-set into not only physics but all of science. His remarkable insights are at the heart of all of thermodynamics. He was, in particular, responsible for interpreting the fundamental origin of entropy—the tendency of a system toward disorder, otherwise known as the second law of thermodynamics—as a function of the ginormous number of possible arrangements of atoms and/or mole-

---

[15] Boltzmann wouldn't have had to wait until the 1980s for the invention of the STM. We've been able to see atoms since the mid-1950s, albeit in much more restricted confines and geometries than are possible with scanning probes. The first image of individual atoms was captured in the autumn of 1955 when Kanwar Bahadur and his PhD supervisor, Erwin Müller, stared at the fluorescent screen of a field ion microscope (FIM) in an otherwise completely dark lab at Penn State University. (This was in the days before image intensifiers. Müller and his colleagues had to allow their eyes twenty minutes or so after walking into the lab to adjust to the darkness in order to discern the faint pattern of atoms on the screen.) For Müller this was a momentous day, representing the culmination of over twenty years of challenging and, at times, immensely frustrating work. According to Bahadur, when Müller saw the image, he declared, "This is it! Atoms, *ja*, atoms!" If only Boltzmann had been in that lab. (Coincidentally, although an FIM forms atomic resolution images using an entirely different method than an STM or AFM, it has one very important component in common: an atomically sharp tip.)

[16] At first glance, it may seem a little odd that in a gas made up of identical atoms (or molecules) there will be a distribution of speeds. The reason that the speeds span a range of values is that, in each collision, the colliding particles exchange energy and momentum. Both the average speed of the molecules and the width of the distribution increase with temperature—pump more heat energy in and the particles can achieve higher speeds.

cules. I mean, you may think that Avogadro's number is big, but it's peanuts compared to the vastly, hugely, mind-bogglingly large number of ways you can *arrange* Avogadro's number of atoms.[17]

In tandem with James Clerk Maxwell—another giant of nineteenth-century physics who, in an astounding intellectual leap, reconciled electricity, magnetism, and light as different aspects of the same fundamental phenomenon (electromagnetism)[18]—Boltzmann determined how the speeds of the atoms in a gas are distributed. In other words, Maxwell and Boltzmann derived a mathematical formula that tells us how many molecules in a gas we can expect to be traveling with speeds between, say, 666 and 667 kilometers per hour (or between 1984 and 1985 km/hr, or 77.7 and 77.8 km/hr or whatever range you choose).[19]

Their mathematical formula is rather sophisticated, but we don't need to write it down to understand the information it provides on the molecular frenzy of an ideal gas. Here's a graph of the Maxwell-Boltzmann distribution:

---

[17] I hope I can be forgiven yet another hat tip here to Mr. Adams. I used to teach a first-year module on introductory thermodynamics (with a smidgeon of statistical mechanics thrown in) and took the opportunity to introduce Boltzmann and the second law of thermodynamics in the context of Adams's infinite improbability drive (from *The Hitchhiker's Guide to the Galaxy*).

[18] You think that'd be more than enough for a lifetime's work, wouldn't you? But Maxwell also made major contributions in astronomy (he was the first to work out the structure of Saturn's rings), the phenomenon of color vision, process control theory, and a number of mathematical techniques that are now second nature to physicists, including dimensional analysis. Oh, and he wrote poetry, too. He almost puts Bruce Dickinson's work ethic to shame.

[19] We can't talk about the probability of a molecule traveling at *precisely* 666 mph for exactly the same reason that, back in Chapter 7, we couldn't have a member of Angle of Death who was *exactly* six feet tall. We can only talk about the probability within a certain range. The probability of traveling at precisely 666 mph is zero because "precisely" in a mathematical sense can only be defined in the context of an infinite number of decimal places of precision.

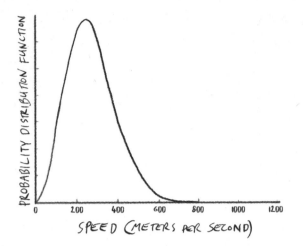

That simple curve distills the frenetic motion of the molecules of an ideal gas down to a representation that's somewhat easier to grasp. It's what's known as a probability density function. You've seen this type of thing before, but in rather different contexts. For one thing, every STM image is also a probability density function: a map showing the probability of finding electrons in a certain region of space. The Maxwell-Boltzmann distribution similarly tells us about probabilities, but it describes the *speeds* of molecules rather than their positions. From the Maxwell-Boltzmann distribution for a gas we can work out the fraction of its molecules traveling between certain speeds. And we do that by slicing up the curve.

Like many parents, I did a lot of hypothetical pizza slicing when I was teaching my kids about fractions.[20] It's much the same principle here: the fraction of molecules traveling between 400 and 500 meters per second can be found simply by taking the relevant slice of the curve and measuring the area of that slice. Alternatively, we can think of that

---

[20] Pizza is, of course, a key food group in the metal community. Has anyone ever written a suitably metalized paean to pizza, you might ask? Well, of course. A quick Google search finds a delightful ditty entitled "Death Metal Pizza." Sample lyric: "Extra large, give it to me now. If you mess up my order you'll be disemboweled." (https://www.youtube.com/watch?v=G3gTBpLHzLY. Lyrics © Jared Dines.)

slice as the probability of finding the molecules traveling between those speeds. Like this:

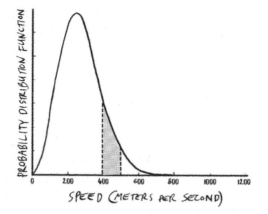

Raise the temperature of the gas by warming it up (inject some heat energy), and the distribution of speeds changes. As you might expect, the higher the temperature, the greater the probability of finding faster moving molecules:

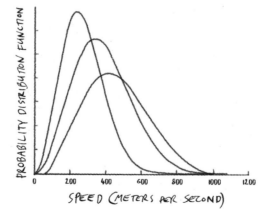

That's how it works for the ideal gas. What's remarkable is that the selfsame mathematical function also accurately describes a mosh pit.

Denim and Leather Brought Us All Together[21]

At first glance, it might not seem so odd that a crowd of moshers behaves much like the molecules in the ideal gas. After all, in both cases, at the most basic level of description, both the gas and the pit are made up of objects frantically rushing around and bumping into each other. But moshers are self-propelled to a much greater extent than the molecules of an ideal gas; they at least have some volition (albeit the extent to which this is true may depend on the degree of consumption of certain fluids prior to or during the concert). Moreover, when moshing metal fans collide, there's a dissipation of energy. This doesn't happen in the ideal gas. The collisions between gas molecules are what physicists call *elastic*. The *total* kinetic energy of the particles is exactly the same before and after the collision, though the kinetic energy (and momentum) of individual particles may change.

A perfectly elastic collision like this is yet another of those idealizations that physics is rife with, another manifestation of the famously frictionless universe that we physicists often prefer to inhabit. (It's perhaps unsurprising that a theoretical model known as "the ideal gas" should be based on substantial simplifications and idealizations.) In the real world, if Metallizer's tour bus were to crash into the wall of their next venue, you'd be very surprised indeed if it simply bounced off soundlessly and went back the way it came without so much as a scratch. That would be an elastic collision. *Inelastic* collisions instead dissipate the kinetic energy of the colliding objects in many ways . . .

---

[21] From Saxon's "Denim and Leather" (*Denim and Leather*. EMI, 1981). Lyrics by Peter Rodney Byford. © Carlin America Inc.

. . . including substantial material damage from the impact, the generation of a loud acoustic impulse (otherwise known as a Big Bang), and the discombobulation of the occupants of the bus.

When moshers collide there's generally less damage than for the bus-meets-wall scenario, but it's nonetheless still an inelastic collision. Moshers don't just bounce, soundlessly, off each other, following a perfect line of motion until their next collision; they're not metaphorical billiard balls or friction-free ideal gas molecules. Yet, ultimately—collectively, statistically—they behave for all the world as if they were. Despite energy being continually redistributed in mosher-to-mosher interactions—"leaking out" of the collisions into the surroundings in a manner that is generally exceptionally difficult to monitor—the speeds in a mosh pit follow a curve that, astoundingly, is the same (within the experimental margin of error) as the Maxwell-Boltzmann distribution. Here's the evidence from Jesse Silverberg et al.'s research:

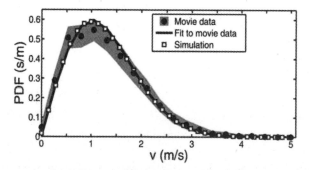

Reprinted with permission from Silverberg, Jesse L., et al. "Collective Motion of Humans in Mosh and Circle Pits at Heavy Metal Concerts." *Phys. Rev. Lett.* 110, 228701. © 2013 American Physical Society. (https://doi.org/10.1103/PhysRevLett.110.228701.)

The moshers' motion produces a distribution of speeds virtually identical to that for a bunch of molecules bouncing around in two dimensions. (The moshers are, of course, moving along 2-D paths; they're not flying around the venue.[22]) But as Jesse et al. point out in the paper, and as I've been at pains to highlight above, there's a big difference in *how* the energy is distributed for the ideal gas described by the Maxwell-Boltzmann distribution and a less-than-ideal collection of moshers at the metal gig.

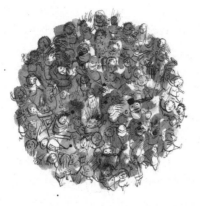

The ideal gas is an example of what's called a system at equilibrium: there's no energy flowing in or out of the group of molecules. Paradoxically, although those nanoscopic billiard balls are flitting back and forth at high speeds, they're also in their equilibrium state. There are many complicated ways of looking at the concept of equilibrium in thermodynamics, but in our case it just means that the average kinetic energy of the molecules, and thus the average temperature, is the same everywhere in the gas. Inject some heat energy, however, and that will no longer be the case: there'll be a region of the gas that will be a bit hotter than the rest. The gas is now *out* of equilibrium. But that hot spot will start to fade away and, after a while, the injection of heat will no longer be localized; the continual colliding of the molecules will have redistributed

---

[22] Having said that, repeating the experiments with jet pack–enabled moshers to check for a 2-D–to–3-D transition in the dynamics would make for a fascinating experiment. That would be a fun grant application to write.

the energy throughout the gas. Equilibrium is restored. Molecules go on colliding elastically, no energy "leaking out" from these collisions, the total kinetic energy of the particles in the gas remaining the same.

Into the Pit

For the mosh pit, however, there's no equilibrium state. It hardly seems surprising to say that a mosher never experiences equilibrium, of course. They're not generally the most chilled-out of individuals, and a search for equilibrium and balance is hardly high on the list of priorities during a metal gig. Leaving aside the New Age connotations, however, and focusing on the physics, it is still the case that a mosh pit is a far-from-equilibrium system. It must be: while collisions in an ideal gas are elastic, collisions between moshers are decidedly not. Energy leaks out in the form of sound, impact with the floor, and so on—there's too much energy emitted from the pit for it to reach equilibrium. Yet not only does the mosh pit study carried out by Jesse et al. show that a mosh pit behaves like a gas of molecules (which is surprising enough in itself), but the results demonstrate that the crowd of moshers, despite being in a highly excited and excitable state, weirdly look as if they *are* at equilibrium. This is a fascinating and counterintuitive result.[23]

In addition to making measurements from publicly available videos, the Cornell team also developed computer models to simulate the behavior of moshers. In another inspired choice of terminology, they coined the acronym MASHer: Mobile Activated Simulated Humanoid.

---

[23] Intuition is a strange, some might even say counterintuitive, aspect of science. Physicists, in particular, are fond of saying that a particular result or observation is "intuitive"—it lives up to our expectations. But just where does that intuition come from? And shouldn't scientists aim to be as disinterested and objective as possible? These are thorny philosophical, psychological, and sociological questions. My colleague Jonathan Tennant here at Nottingham is one of a number of philosophers who are keen to unravel the intricacies of intuition in science. *Sixty Symbols* featured Jon in this 2014 video: www.youtube.com/watch?v=0jRppd6c5J8.

The MASHers in their simulation were set up to have the following characteristics: soft-body repulsion, self-propulsion, and flocking interactions.

"Soft-body repulsion" is not a comment on the physique or self-image of the typical mosher. Instead, it's a physics term that refers to how the mosher/MASHer particles deform and bounce off each other in the pit.

"Self-propulsion" is, I hope, self-explanatory.

The "flocking interactions" part of the model is key to capturing the dynamics of moshing. Each of the moshers/MASHers can be associated with a vector describing its velocity. A vector is just a little arrow whose length in this case represents the speed of the MASHer. The direction the vector is pointing tells us which way the mosher is headed at a particular instant of time. The flocking interactions are determined by taking into account all of the velocity vectors within a certain radius of a given MASHer: the crowd response in the vicinity of a mosher—like, say, our friend Marsha—therefore contributes to determining her motion.

The moshing simulation is set up so that there are both passive and active MASHers (as shown in the sketch above). Those simulated humanoids involved in active mashing are self-propelled and subject to flocking behavior, while passive MASHers prefer to remain stationary and watch as the madness around them unfolds.

One final ingredient in the mix of parameters was required to accurately model moshing, and I don't think you'll be too surprised to hear just what's missing.

Noise.

But this is not noise in a *"Bring the . . ."* or *"C'mon, feel the . . ."* sense; it's got nothing to do with the decibels flowing from the stage. This is noise in its broader scientific definition. Mention noise to a physicist—even a metal-loving physicist—and a frustrating experience in the lab is rather more likely to spring to mind than the title of that genre-defining Public Enemy–Anthrax collaboration. To a physicist, noise is unwanted "buzz" that contaminates a signal. A great deal of experimental science revolves around the tedious and exasperating process of isolating a signal from noise pollution either via very careful experimental design or judicious filtering. Or very often both.

The potential sources of noise in an experiment are wide, varied, and deeply irritating. Often noise is due to electrical contamination. For example, scanning probe microscopists measure very-low-level electrical signals (sometimes down to just femtoamps—i.e., 10^{-15} amps—of current), and we fight against noise that can arise not from the instruments and equipment in our labs but that in other rooms in the buildings in which we're based. Often electrical noise will have a well-defined frequency of 50 Hz and/or some related harmonics—100 Hz, 150 Hz, 200 Hz, etc. This stems from the frequency of the AC supply in the UK. (Our North American friends will instead suffer with 60 Hz noise.)

The most effective way to deal with noise of this type is to kill it at its source. But finding the source can often be akin to hunting down that mythical pot of gold at the end of the rainbow. In the dark. With no sign of the morning coming. Days, weeks, or even months can be lost in increasingly fraught hunting expeditions chasing that elusive noise source. In the event that the origin of the noise can't be snuffed out, an alternative (though less preferred) strategy is to filter it out. Here's a visual representation of what I mean:

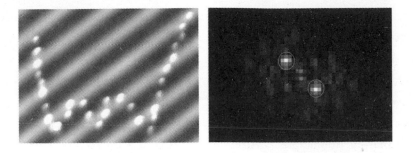

On the left, the "signal"—our universal metal symbol written in atoms—is contaminated with a single-frequency noise source (somewhat similar to what happens when we have contamination from the AC supply in an experiment). The corresponding Fourier transform of the noisy image is shown on the right. The noise appears as bright spikes (highlighted with circles).[24] If we filter out that specific frequency, here's what we get . . .

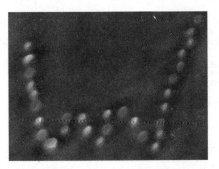

---

[24] "But why two spikes? It's a single frequency, right? So shouldn't that be a single spike?" Well spotted. There are two spikes for a single frequency due to the traditional way in which 2-D Fourier transforms are plotted. As described in the Appendix, there's a deep link between trigonometric and exponential functions. A sine function can be written as follows: $\sin(\omega t) = e^{i\omega t} + e^{-i\omega t})/2i$, where i is $\sqrt{-1}$. Note that there are two exponential functions making up the sine, one with a positive frequency and the other with a negative frequency. This is what gives rise to the two spikes in the Fourier transform: there's a spike at positive frequency, ω, and another spike at negative frequency, $-\omega$. The negative frequencies are simply a "mirror" of the positive frequencies—there's no additional information there—so, in principle I could have just shown you one half of the Fourier transform. But this is a book for metal fans, and I know how much you love the gory detail . . .

That's not too bad, is it? In some cases, this filtering strategy is good enough, especially if the noise is "narrow band" (only a very small range of frequencies[25]). But often the noise is broadband, spanning many frequencies. Just as white light comprises all the colors of the visible spectrum—a wide range of optical frequencies—the term *white noise* is used to describe the combination of a similarly wide variety of contaminating frequencies. Instead of being concentrated at a particular frequency, as in the example on the previous page, white noise spans a vast spectrum of frequencies. Here's our image of the atomic horns with the addition of white noise—alongside its Fourier transform:

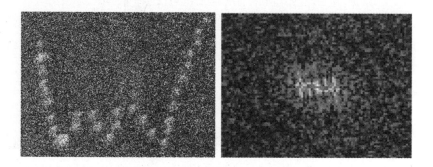

This now represents rather more of a challenge in terms of filtering. We can't simply select a narrow band of frequencies and chop those out; instead, the noise is distributed over the entire frequency space, mixed in along with the true image information.

We've used a visual example, but of course we can also listen to the sound of white noise, as Anthrax informed us all the way back in 1993.

---

[25] Remember Dirac's delta: A perfect single-frequency signal only exists in the universe of pure mathematics. It can't exist in the real world because that would imply a sine wave with an infinite duration. Any real signal always has a finite bandwidth. That width may be so small that we can't measure it, but it's nonetheless always there. This is once again the uncertainty principle in action.

Here's what the frequency spectrum of white noise, of the type that opens up that classic Anthrax album, looks like:

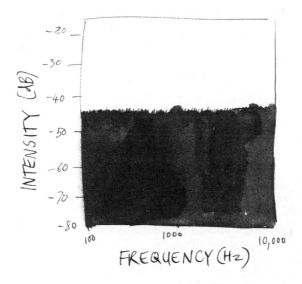

Note the flatness of the spectrum: all noise frequencies are equally present (give or take) and equally irksome. This is the definition of a white noise source. (There's also pink, and blue, and red noise. In each case, the label comes from drawing a parallel between the frequency spectrum of the noise and of light of that particular color.)

The key thing about white noise is that it's random. There's no signature frequency (or set of frequencies) there. This in turn means that there's no relationship between the volume of the signal at a particular time and that at any other time. Or, to switch back to a visual rather than an audible signal, pure white noise means that the brightness of a particular pixel in an image due to that noise is random and so tells us nothing about the brightness of its neighboring pixels: if the noise is high at one pixel, the noise at a neighboring pixel might be either high or low. This is what happened with the Metallizer logo: it got buried in a fuzz of pixels. Physicists say that the noise is *uncorrelated*.

This type of uncorrelated noise crops up everywhere in nature. It's at the core of the ideal gas, for one thing. The gazillions of collisions mean that each individual molecule's motion is scrambled: it's not correlated with what its neighbor might be doing. If we could somehow capture each and every pressure fluctuation when a molecule strikes a wall of the surrounding container and convert it into a click or a beep, we'd hear something not entirely different from those opening few seconds of *Sound of White Noise*.

From one perspective, it's quite straightforward to introduce noise into a computer simulation: a random number generator of some type can be used to throw the equivalent of a (multisided) digital die, and the resulting number is then fed into the model. Computational scientists, however, can lose a great deal of sleep over the question of just how random their computer's random number generator really is. Technically, it's only possible to generate *pseudo*-random numbers on a computer (because the algorithms that are used are, at their core, most definitely not random). Often, however, pseudo-random is random enough. In Jesse et al.'s case, their noise-generating function was tried and tested and was more than up to the task of modeling moshing.

Bringing the noise was key to the MASHer simulation's ability to accurately capture the motion of moshers. When the noise is increased so that it dominates over the flocking tendency—or, in other words, when the MASHers act as individuals, bouncing around randomly without being influenced by their peers—the motion of active MASHers in the simulation not only looks like what happens during metal gigs, but the numbers and curves agree with those measured directly from real-life videos of mosh pits. Or, in the rather more staid language of science, there's both qualitative and quantitative agreement between theory and observation.

If, instead, the tendency to flock is increased so that it dominates the noise in the simulation, a remarkable effect occurs: a vortex-like state is produced where active MASHers phase separate to form a collective motion with nonzero angular momentum. It looks like this:

I suspect you'll have guessed by now that there's a less technical term to describe this nonzero angular momentum state: it is, of course, a circle pit. Once again, the MASHer model, which reduces all of the human psychology and sociology at play in the pit down to a simple set of rules of motion, does an exceptionally good job of re-creating the dynamics. One fascinating aspect of the circle pit not captured by the MASHer simulation, however, was the direction of the motion. In the simulations, it's 50:50 counterclockwise vs. clockwise rotation in the pit.

But not in the real world. Video footage from a wide variety of gigs shows that, remarkably, 95 percent of all circle pits flow in the counterclockwise direction. Before you ask, this strong preference for counterclockwise motion doesn't depend on whether the gig happened in the northern or southern hemisphere;[26] indeed, there was no dependence on geographical location at all. The Cornell researchers don't have a definitive explanation for the rotational "anomaly" but make the very plausible suggestion that it points to the much larger percentage of right-handed people among the world population. Once again, we find

---

[26] The discovery of a metal-specific manifestation of the Coriolis effect would have certainly shaken up physics as we know it.

that symmetry informs behavior. The prevalence of right-handedness will set up a rotational asymmetry in the pit because a mosher will have a preference to set off on their circuit on either their right or left foot.

Beyond the Walls of Death

The Cornell paper on moshing deservedly attracted quite a bit of media attention. This led, of course, to a smattering of the usual sniffy and disgruntled comments from those who lurk below the line: "Taxpayers' money wasted on frivolity. Why aren't these boffins curing cancer instead of spending their time, facilities, and faculties analyzing rock concerts?"[27]

Leaving aside the minor issue of physicists' distinct lack of training in the advanced biomedical techniques and protocols required for cancer research, the insights on moshing behavior provided by Jesse L. Silverberg and his colleagues do, in fact, have much broader societal significance. As they put it in their paper: "Further studies in this unique environment may enhance our understanding of collective motion in riots, protests, and panicked crowds, leading to new architectural safety design principles that limit the risk of injury at extreme social gatherings." Indeed, this is exactly where the moshing study led: Silverberg is currently examining how crowds behave in situations such as the nadir of consumer culture that is Black Friday (which makes Megadeth's frantic song of the same name seem positively tame by comparison).

We started this chapter with a consideration of community in the metal subculture; it seems fitting to close, as we have, by drawing connections between the metallosphere and the none-less-metal environs

---

[27] I love the line at the end of the Acknowledgments section of Jesse et al.'s paper: "Fieldwork was independently funded by JLS."

of a shopping mall.[28] Indeed, considering those broader connections would seem to be the most appropriate way to draw this entire book to a close. We've seen throughout just how metal's iconography, culture, and, of course, music are intimately connected with many of the most fundamental and esoteric concepts in physics. In the conclusion—and in the traditional spirit of the concluding sections of a scientific paper (or PhD thesis)—I'm going to take stock of what we've covered and suggest some possible avenues for further research at the tantalizingly tangled intersection of metal, physics, and society.

---

[28] Niggling at me while I write this stuff is the concern that my words will be ripped out of context and end up being featured someday in the "Pseuds Corner" section of the wonderful *Private Eye*. (If you're not familiar with the superlative satire of the UK's *Eye*, might I recommend you subscribe ASAP? You can, of course, cancel your subscription forthwith should the content offend.) See http://www.private-eye.co.uk/.

Conclusion

AND THE BANDS PLAYED ON . . .

See the people, feel the power
There were sixty thousand there
Just like thunder, the crowds began to roar
Were you there? Did you know?
Did you see all the show?
There was magic in the air

—from Saxon's "And the Bands Played On"[1]

---

[1] *Denim and Leather*, EMI, 1981. Lyrics by Peter Rodney Byford. © Carlin America Inc.

A few hundred pages later, I'm hoping that I've gone some way at least to convincing you of the many parallels between metal music and quantum physics. This book had its origins in a post I wrote for the Institute of Physics' *physicsfocus* blog (now sadly defunct) four years ago. The post in question had the naggingly familiar title of "When the uncertainty principle goes up to 11" and focused on the relationships between guitar chugs, frequency spectra, and the uncertainty principle—relationships we covered (in substantially more depth) in Chapter 7. After that post was uploaded, I idly wondered for a moment or two whether there might be some mileage in expanding on the theme in something weightier than an online piece, drawing out the broader links between metal and quantum physics. At the time, I came to the conclusion that while my "metal meets quantum" idea might work for a one-off blog post, it'd be a bit of a stretch to extend it to a full book. I put the concept to bed and got back to work on other things.

But I couldn't let it rest. I was particularly tickled by the notion of juxtaposing quantum physics—not the *least* cerebrally challenging of subjects—with heavy metal, which, as we've discussed elsewhere in these pages, is not generally considered the most intellectual of genres. And the more I thought about it, the more connections I saw, and the more a marriage of metal and (mathematical) physics made sense. So when I subsequently received an email out of the blue from a literary agent asking whether I had any ideas for a book, I suggested that the parallels between metal music and quantum physics might be a topic worth exploring. I was expecting a response along the lines of, "Errmm, okay. Thanks for that. Don't call us, we'll call you." So I was exceptionally surprised when not only was a potential publisher, the wonderful BenBella Books, interested in hearing more, but they in turn offered me a contract for the very book you've just (nearly) finished reading.

Having bitten the bullet and written the thing, I can now say that the primary difficulty I encountered in writing *When the Uncertainty Principle Goes to Eleven* was not, as I mistakenly first thought, a struggle to find material because the theme was limited in scope. Instead, I repeatedly had to constrain myself from introducing yet another metal-physics crossover topic. I was spoiled for choice. And then some.

Given this embarrassment of riches, you might be wondering: Which aspects of the metal-physics nexus did I leave on the cutting-room floor? If we were to go one louder again—if we could turn up to an unprecedented, ear-shattering twelve—which facets of metal would I turn my focus to, or explore more deeply? And what perspectives would we gain?

The guitar is a wonderfully expressive instrument, and it's exceptionally difficult to do justice to the physics and psychology of that expressiveness in a relatively short book. My fellow Dublin City University alumnus David Robert Grimes (currently a research fellow at University of Oxford) published a fascinating study of the physics of lead guitar a couple of years ago, in which he developed mathematical models for string bending, vibrato, and whammy bar dynamics.[2] Could we go even further? Could quantum physics provide insights into vibrato, for example? According to a research team at Queen Mary University of London,[3] the answer is a resounding yes. Moreover, the first words of the title of Grimes's paper—"String Theory"—highlight another rich seam of metal-physics links to be tapped. Being a nanoscientist, in this book I stopped short at the atomic, (sub)molecular, and nanoscopic levels of matter, but, as the physicist and talented guitarist

[2] Grimes, David Robert. "String Theory: The Physics of String-Bending and Other Electric Guitar Techniques." *PLOS One* (2014). doi:org/10.1371/journal.pone.0102088. This is an open-access paper, so free to download at: journals.plos.org/plosone/article?id=10.1371/journal.pone.0102088.
[3] "Quantum physics offers insight into music expressivity." *Queen Mary University of London News Page* (March 15, 2017). www.qmul.ac.uk/media/news/items/se/193712.html.

Mark Lewney highlighted in a series of lectures,[4] some of the more stringy aspects of particle physics and cosmology can also be explained via guitar analogies.

I could have dedicated an entire chapter to distortion, nonlinearities, and feedback, but given that I was already ten thousand words over the limit when I submitted the manuscript, I suspect that my exceptionally patient editor would have felt that was overkill. Nonetheless, feedback has not only been at the heart of rock and metal since Hendrix's incendiary playing introduced the technique, it's integral to many aspects of physics—at all scales—including, in particular, chaos theory. I am firmly of the opinion that a great deal of physics could be taught through the medium of metal amplification. (It would be one way of holding students' attention, at least.)

One of my favorite-ever scientific papers—certainly in my top ten—was published by Hari Manoharan and his colleagues at Stanford University and has the rather technical title of "Quantum Phase Extraction in Isospectral Electronic Nanostructures."[5] This belies the elegance of the research and disguises its close links to another core element of metal that didn't get enough attention in this book: drumming. Manoharan et al. addressed an age-old question with their research: Can we hear the shape of a drum? But they did it at the quantum level, building nanoscopic drum heads by manipulating single molecules one at a time. It's a hugely impressive piece of work that brings together maths, music, and physics at the most fundamental level. Maybe next time.

The snap of a snare and the thud of a bass drum are clearly not melodic, sustained sounds like those produced by a resonating guitar string; they're better described as impulses. There are entire branches of

---

[4] "Where Physics and Guitars Collide." The Institute of Physics Schools and Colleges Lecture 2008 (published March 3, 2011): www.youtube.com/watch?v=fjKugarLN-s. (But remember the golden rule of YouTube viewing: *Don't read the comments.*)

[5] See Moon, C. R. et al. *Science* 319, no. 5864 (2008): 782–7.

physics, engineering, and maths dedicated to impulse response theory that I feel somewhat guilty for not getting round to discussing, especially as a key aspect of those areas is the work of a humble mathematical physicist known as George Green. Green, of "Green's function" fame (well, among scientists, mathematicians, and engineers), is very closely connected with Nottingham, as he owned and ran a mill in Sneinton, a suburb of the city, in the nineteenth century. As I write this chapter, I can look out the window of my office and see the George Green library beside the Physics building. Like Fourier, Green's name is synonymous with an exceptionally important mathematical approach to gaining deeper insights into how our universe behaves.

Sticking with drums, and returning to the theme of chaos theory, there's a wealth of physics and maths to be found in beats, from the most delicate and subtle of Neil Peart's cymbal accents to the pounding double bass attack of Dave Lombardo. A wonderful quote from the polymath Gottfried Wilhelm Leibniz—a contemporary of Isaac Newton who, to Newton's famed chagrin, independently developed calculus—neatly captures the close interrelation of numbers and rhythms: "Music is the pleasure the human mind experiences from counting without being aware that it is counting."

Embedded in the fluctuations of drum rhythms are deep connections with noise in natural systems, ranging from the timing of heartbeats to the sound of raindrops striking a surface. A number of talented undergrads here at the University of Nottingham have carried out a detailed analysis of the variations in the sixteenth note hi-hat pattern of Rush's "Tom Sawyer,"[6] based on the approach by Esa Räsänen and his colleagues.[7] We'll write these data up and submit the paper soon,

---

[6] You didn't think I'd close the book without including just one more reference to Messrs. Lee, Lifeson, and Peart, did you?

[7] Räsänen, Esa et al. "Fluctuations of Hi-Hat Timing and Dynamics in a Virtuoso Drum Track of a Popular Music Recording." *PLOS One* (2015): doi: org/10.1371/journal.pone.0127902. This is an open-access paper, so free to download at journals.plos.org/plosone/article?id=10.1371/journal.pone.0127902.

but suffice to say that a study based around the analysis of one of my favorite-ever songs was not one of the most tedious pieces of research I've been involved with over the years.

Less frivolously, the drum correlations in Rush's (arguably) most famous work once again highlight those strong maths-music-science-art links I've been hawking throughout this book. Hopefully, society will eventually evolve to the point where these needless interdisciplinary divides are seen for precisely what they are: irksome limitations to understanding and to broadening our knowledge. STEM students can learn a great deal from the arts and humanities, and vice versa. It's a great shame that some can't see beyond the narrow confines of their chosen discipline and prefer instead to exist within what might fashionably be called an echo chamber.

In the end, let's just say that we've barely scratched the surface—there's so much left to learn from metal music and physics alike, so many links yet to be made. In fact, after careful consideration, I'm confident that an entire university course could be developed that explains physics through the medium of metal. Before those learned academics who may be reading wrinkle their noses and run to the hills at the very suggestion, I'm entirely serious. In fact, I'll go further. Why not make that course truly interdisciplinary in scope? Why not combine physics, psychology, sociology, art, and maths in the analysis of metal? We could engage and excite students about scientific principles and theories and, in parallel, address questions about the development of subcultures, iconography, in-group vs. out-group dynamics, history, and the social norms associated with a major genre of music.

Nottingham Trent University—we're fortunate to have two universities in this fair city; I work at the other one—has already paved the way for a course like this, having accredited a Foundation Degree in Heavy Metal Performance offered by New College Nottingham. The introduction of the course led, of course, to predictable outrage about lack of rigor and the growth of "Mickey Mouse" degrees. But while I can understand how the idea of a degree in heavy metal might be a hard sell for many employers, engagement is *everything* in education. And what better way to ensure that students are thoroughly engaged than to ground what they're learning in a music and (sub)culture they love?

I'm not the only academic thinking along these lines. In the fall of 2014, and with none of the bombast one might have reasonably expected, the first issue of a new scholarly journal entitled *Metal Music Studies* was published.[8] Could this be the first step toward academia accepting the educational power and potential of metal? Perhaps. Here's how the editors, Karl Spracklen and Niall Scott, of Leeds Beckett University and the University of Central Lancashire, respectively, described the motivation and future directions of their journal:

> *There is such a diverse range of things expressed through the music and in its culture. It is a rich and varied spectrum that illuminates features of the human condition that very few forms of artistic and popular culture expression touch on . . .* Metal Music Studies *also stands to critique traditional academic approaches to study because like metal itself, it is interdisciplinary . . . The study of metal has a darkly bright future indeed.*[9]

So watch this space. Keep your ear to the ground. As Mr. Halford and colleagues would (almost) say . . .

You've got another think coming.

---

[8] By a publishing house called Intellect. I rest my case.
[9] Invisible Oranges Staff. "New Journal Takes Academic Look at Metal." *Invisible Oranges*, January 28, 2015. www.invisibleoranges.com/new-journal-takes-academic-look-at-metal/.

Appendix

THE MATHS OF METAL

Whether you've read the whole book before arriving here or are dipping into the appendix every now and then to explore a reference in the text, I suspect that you're fully aware you're not going to find an exhaustive, blow-by-blow account of all the mathematics underpinning Fourier analysis, Dirac delta functions, spherical harmonics, and so on. Here we'll delve a little further into the mathematics of metal, but it is more of a snorkel than a deep ocean dive, so to speak. For one thing, *When the Uncertainty Principle Goes to Eleven* is hardly intended as a core text for university physics or maths courses. It could, however, serve as additional reading for introductions to Fourier analysis and quantum mechanics, and it's in that spirit that this appendix is written.

A little of what follows has been adapted from a set of notes I pre-pared far back in the mists of time (well, a decade ago at this writing[1]) for "Applications of Fourier Analysis," a class on, um, the applications of Fourier analysis.[2]

Signals and Sines

The fundamental premise of Fourier's approach is that we can write a function (aka a signal, aka a waveform) as a sum of sines with different amplitudes, frequencies, and phases. If we're dealing with a regularly repeating, periodic function, like an *infinitely* sustained guitar note,[3] we work with a *Fourier series*. Later in this appendix we'll deal with the *Fourier transform* approach, which can be used to treat both periodic and aperiodic (nonrepeating) functions, and which is what we largely focused on throughout the book. (Although not every mathematical function may be represented as a Fourier series or transform, the vast majority of functions of interest in the physical sciences—and, equally importantly for our purposes, in metal—are amenable to Fourier analysis.) While a Fourier transform can do everything that a Fourier series does, the latter makes a handy stepping-stone on the way to developing an understanding of those crucially important time/frequency (and space/reciprocal space) relations.

There are a number of ways of writing the mathematical expression for a Fourier series. We'll stick with one of the most common:

---

[1] The year 2007 was not particularly auspicious for metal releases, although it was noteworthy for the re-formation of such classic acts as Carcass, Sacred Reich, Extreme (settle down there at the back—Extreme had lots of hair, virtuoso guitar solos, and riffs aplenty so they're metal in my book), and, of course, the mighty Tap.

[2] In case you're interested, the notes are still available here: www.nottingham. ac.uk/~ppzpjm/F32SMS/F32SMS%20Notes%20Set%201%20%2706-%2707. pdf.

[3] . . . because if it isn't running to infinity, then it isn't regularly repeating.

$$f(t) = A_{DC} + \sum_{n=1}^{\infty} [A_n \cos(n\omega_0 t) + B_n \sin(n\omega_0 t)]$$

Let's translate the maths above to English. First, the t symbol represents time throughout. That equation is telling us that a periodic function, which we're calling $f(t)$, can be represented by adding a constant (a number like 1 or 2 or 21.12 or 742617000027 or . . .), which we've called A_{DC} for reasons I'll get to, to a lot of sine and cosine functions. Really, a *lot* of sine and cosine functions. The only difference between a sine and a cosine function is a phase shift. (See Chapter 4.) Shift a sine function by $\pi/2$ radians and you've got a cosine. Throughout the book I've focused on sines because, when it comes to explaining the concept of Fourier analysis, the additional complication of considering cosines isn't really worth the effort. (Really, it's not.) But here in the appendix, we can afford to be a little more rigorous. (Stop scoffing at the back there, you mathematicians. This is what counts for rigor among experimental physicists.) The traditional approach to introducing Fourier series takes into account both sines and cosines, so it'd be a little remiss of me not to incorporate sine's humble sibling function here.

As you may be aware, the Σ symbol on the right-hand side (otherwise known as "sigma") means "the sum of." In a world of pure mathematics, it makes sense to sum up an infinite number of terms—we can just keep adding terms until we get the ideal representation.[4] (Usually.) That's what we're doing here, as indicated by the infinity sign on top of that Σ symbol. The $n = 1$ on the bottom means that our infinite set of terms starts by setting n equal to 1 (then we'll step up in increments of 1 . . . and keep going).

---

[4] This type of infinite summation isn't possible in the real world. For one thing, we can only ever have an approximation in the real world because we only ever have a finite amount of time at our disposal. We have to be satisfied with adding a finite number of sine waves, continuing until we get close enough to the function we want to represent. But then, "close enough" is generally good enough for a physicist (spherical cows, frictionless surfaces, infinite sustain . . .). And even sometimes an infinite number of terms isn't enough either. But that point really will have to wait for the sequel . . .

So, the first two terms in the series represented by the Σ symbol would be as follows:

$$A_1\cos(\omega_0 t) + B_1\sin(\omega_0 t) + A_2\cos(2\omega_0 t) + B_2\sin(2\omega_0 t)$$

Again, in English: we're just adding up sine and cosine waves with different frequencies (ω_0 or $2\omega_0$ in this case because $n = 1$ and then $n = 2$). The constants A_1, B_1, A_2, and B_2 (like A_{DC}) just represent simple numbers—they have no dependence on time. They're boringly and unflinchingly constant. But, importantly, they tell us just how much of each frequency component we need in the "mix" to represent our function $f(t)$; they tell us the amplitudes of the waves that are required to build our function. So we end up with a frequency spectrum that comprises a series of spikes, like this:

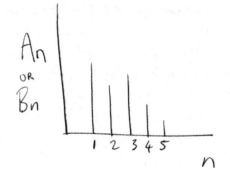

Now, I'm obviously not going to write out all the terms in that infinite series represented by the symbol below:

$$\sum_{n=1}^{\infty}$$

as that would be almost, though not quite, as tedious as listening to the entirety of Yes's *Tales from Topographic Oceans*.[5] But I hope you get the idea. We're just adding up sines (and cosines) to build a pattern, $f(t)$. In

---

[5] I'm a big prog rock fan but, c'mon, *Tales* . . . is just beyond the pale.

this case that pattern is in time, but as we see elsewhere in the book, the same arguments hold equally well for patterns in space. (If it's not too traumatizing at this stage, you might think of an infinitely long pair of those striped Stryper strides . . .)

As we also see many times throughout the book, a sine or a cosine function oscillates back and forth so that its average value is zero—it spends as much of its time being positive as it does being negative. But many functions don't have an average value of zero. So we need a way to take account of this. And that's where A_{DC} comes in. It allows us to take account of any offset, like this:

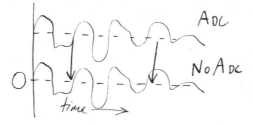

In other words, all that A_{DC} does is shift the function up or down the y-axis. If A_{DC} has a value of 0, then the function has an average value of 0. Any non-zero value of A_{DC} will mean that the function (or waveform, or signal, or sample, or whatever we want to call it) will be pushed away from the x-axis (which represents $f(t) = 0$).[6] A reasonable analogy is to think of the function as being bumpy terrain. If the terrain is close to sea level (in terms of its elevation) then the value of A_{DC} is close to zero. If, instead, that terrain is 500 meters above sea level, then we take the increase in elevation into consideration by adding an A_{DC} term (in that case, with a value of 500 meters).

---

[6] For those who are curious, the "DC" in A_{DC} stands for "direct current." It's borrowed from the language of electrical engineering. A "DC" signal is boringly flat and constant. By contrast, an "AC" signal (like a sine wave, for example) *alternates.*

The Most Beautiful Equation?

Mathematics, rightly viewed, possesses not only truth, but supreme beauty—a beauty cold and austere, like that of sculpture, without appeal to any part of our weaker nature, without the gorgeous trappings of painting or music, yet sublimely pure, and capable of a stern perfection such as only the greatest art can show.

—Bertrand Russell (1872–1970), from *A History of Western Philosophy*

Before we can make the leap from the Fourier series approach described in the preceding section to the broader concept of a Fourier *transform*, we need to take a look at the fundamental connection between sines and cosines (i.e., trigonometric functions) and exponential functions. This is best captured in what has been regularly described as the most beautiful equation (or theorem) in mathematics: Euler's identity. Here's what it looks like:

$$e^{i\pi} + 1 = 0$$

This formula links five fundamental constants: the numbers 0 (otherwise known as the additive identity—adding zero to a number returns the same number), 1 (the multiplicative identity), π, e (the exponential constant that crops up all over science, from probability theory to, as we'll see, the analysis of sound waves and signals), and the imaginary number, i, the square root of -1. Whether you feel that this identity of Euler's deserves its reputation as the most beautiful equation in mathematics, or lives up to Russell's "cold and austere" definition of mathematical beauty, is entirely a matter of taste. There is a feeling among some—slightly grumpy, it must be said—groups of physicists and mathematicians that Euler's identity is rather overhyped. But then, beauty in all things is in the eye of the beholder.[7]

That's all well and good, you might say, but there doesn't appear to be a sine or cosine anywhere in the vicinity of that equation. Where's the link between trigonometric and exponential functions I promised?

---

[7] "Do you see what I see?"

Euler's identity is, in fact, a specific expression of the Euler formula, which makes very clear the close connection between sines, cosines, exponential functions, and the imaginary number i:

$$e^{i\theta} = \cos\theta + i\sin\theta$$

I'm not going to prove that equation here. The Wikipedia page for the Euler formula is well worth a read and outlines two approaches to the proof, involving either series expansion or calculus. For our purposes, all you need to appreciate is that Euler's formula provides us with that key link between exponential and trigonometric functions (via complex number analysis).

Infinite Schemes

So, we've already outlined the maths of how we break down a regularly repeating pattern/function/waveform: we sum up a series of sines and cosines that have a discrete set of frequencies. But there are lots of functions and patterns that don't regularly repeat. (Although some metal bands' vocals and riffs might be a tad repetitive, they don't actually just repeat exactly the same thing over and over again, *ad infinitum*.[8])

So how do we treat patterns that don't regularly repeat in time with clockwork precision, like a cymbal "stab," or a patented Celtic Frost "death grunt," or a Lars Ulrich backbeat?[9] Yet again, it all boils down to the conjugate nature of time and frequency: a longer period in time is associated with a lower frequency and vice versa. In the Fourier series approach covered above, there's a well-defined period, T, for a regular, periodic waveform—think of a sustained guitar note—which means that there is also a well-defined frequency. The upshot of this is that a periodic function (aka waveform, aka signal) can be broken down into a set of waves that have specific, well-defined frequencies and thus the

---

[8] Well, with the notable exception of Disturbed: www.youtube.com/watch?v=66gSvNeqevg. "One band. One album. One never-ending song."

[9] Sorry, couldn't resist. I've actually got a lot of time for Mr. Ulrich's drumming. See Chapter 3.

frequency spectrum, as sketched on page 314, comprises definite spikes separated by definite spacing.

But what if we have a non-periodic function? Let's take the example of a single, isolated guitar chug. We crunch the guitar just once. We have an entirely aperiodic function (as shown below). But, mathematically, this is in essence equivalent to saying that the period is infinite (the chug never repeats, so to hear it repeat we'd have to wait an infinite period):

And if the period is infinite, then the spacing of the frequency components will be the inverse of infinity: an *infinitesimal* separation. In other words, we move from a set of discrete frequencies for a periodic function to a *continuous* spectrum for an aperiodic function. And that in turn means that we change from a summation to an integral when we're calculating the frequency spectrum.[10] Taking this, and the relationship between trigonometric and exponential functions revealed by Euler into account, we arrive at the formula for a Fourier transform (my favorite mathematical equation):

$$F(\omega) = \frac{1}{\sqrt{2\pi}} \int_{-\infty}^{+\infty} f(t) e^{-i\omega t} dt$$

---

[10] This is not at all obvious, despite my breezy mention of summation vs. integration. I don't want to provide a refresher course on integral calculus here, however—I recommend, for one, Khan Academy for that—but although integration indeed involves summing up lots of bits, those bits come in infinitesimal, rather than discrete, chunks.

You don't need to understand the minutiae of that equation. For one thing, don't worry about the constant $1/\sqrt{2\pi}$ at the start—that's just a scaling factor to ensure that the time and frequency representations are consistent with each other. It's the stuff on the right-hand side of the integral sign that's important. In the context of this book, nothing really matters except the following: the formula for a Fourier transform lets us take a function that varies in time, $f(t)$, and convert it to a frequency spectrum $F(\omega)$. The mathematical conversion process involves multiplying the function by sine and cosine functions (via the $e^{-i\omega t}$ term) and integrating across *all time* (that's why the integral has limits of $+\infty$ and $-\infty$).

When we leave the universe of pure mathematics, however, and get into the messier and more imprecise world of the experimental physicist, we have to accept that we can't handle infinitesimal quantities and infinities. An infinitesimal quantity is an idealization—it can never exist in the real world because we can never make measurements to infinite precision. And when a computer calculates a Fourier transform—and the vast majority of Fourier transforms are calculated using computers—it can't integrate for an infinite time. Even if it could, we'd never get an answer. Infinity is an awfully long time away. So approximations have to be made.

This sounds like we're cutting corners but, really, we're not. Any *real* signal will not only last for a finite amount of time but it will, in the vast majority of cases nowadays, have been sampled by a computer. Just as a computer can't measure for infinite time, it also can't measure infinitely quickly. This means that the signal will have been sampled at discrete time intervals—a computer can't work with infinitesimals because it can't process at infinite speed. Why calculate beyond what we need to calculate? It's a question of knowing our limits and working with what we have.

So, on a computer, the equation for a Fourier transform we saw above is, er, transformed . . . into this:

$$F_k = \sum_{n=0}^{N-1} f_n e^{-nk2\pi i}$$

Again, what's important here isn't understanding the fine detail but the bigger picture. What that equation does is sum up a set of samples of a signal, f_n (which could be anything from a recording of a Meshuggah riff to measurements of the position variations of a mosher in a pit, and far beyond). We write the signal as f_n rather than $f(t)$ now because we have a set of samples at discrete instances of time rather than a continuous function. It's just a series of numbers: f_1, f_2, f_3, and so on. Similarly we write F_k rather than $F(\omega)$ because the frequency spectrum will also be discrete: F_1, F_2, F_3, and so on. We're back to a summation sign rather than an integral because of the discrete nature of the signals, and finally, we can only work with the total number of samples we have, N, rather than an infinitely long signal.

Why Does Kinetic Energy Depend on Velocity Squared?

Back in Chapter 2, I promised I'd explain why the formula for kinetic energy is ½ mv^2 instead of ½ mv. Why is the velocity squared? As this is a maths-focused appendix, let's begin by doing the maths. Or, more specifically, the calculus.

In Chapter 2 we were considering a CD falling from a height, but any object would work—Metallizer's tour bus falling off a cliff (again) would be another appropriate example. In both cases we're converting potential energy into kinetic energy. Putting the mathematical flesh on the bare bones of that statement, we'll start by defining the energy gained by a falling object (or any object subject to any force—it doesn't necessarily have to be gravity) in as general a fashion as possible.

The increase in kinetic energy is equivalent to the work done by the falling object. Now, usually, work done is defined as force multiplied by distance. That's okay as a broad statement, but we can be more mathematically rigorous:

$$W = \Delta E_{kin} = \int_0^d F(z)dz$$

Here W represents the work done by the falling object, ΔE_{kin} is the change in kinetic energy, d represents the distance through which the object falls, and $F(z)$ represents the force—which can vary with position above the surface, z.

But a force can be expressed as mass multiplied by acceleration, so that means we can write the equation above as:

$$\Delta E_{kin} = m\int_0^d a(z)dz$$

where m represents mass and $a(z)$ represents the acceleration, which can again vary with position, z, above the surface.

Now we'll introduce a little mathematical trick to help us get to the answer we require—we'll switch variables.

$$\Delta E_{kin} = m\int_0^T a(t)\frac{dz}{dt}dt$$

We do this type of thing a lot in physics. Note how we've changed from an integral involving movement in space (i.e., in the z direction) to considering the motion of the object as a function of time. T represents the total amount of time the object spends falling until it hits the floor, and $a(t)$ is now the variation of the acceleration in time rather than in space. To take account of this change in variable, our integration must also be in terms of time rather than space, so we write $\frac{dz}{dt}dt$ to account for this. But $\frac{dz}{dt}$ is the rate of change of the displacement of the object with respect to time, and this is just its velocity. Thus, we can write:

$$\Delta E_{kin} = m\int_0^T a(t)v(t)dt$$

And acceleration is just the time derivative of velocity, so we in turn can write:

$$\Delta E_{kin} = m \int_0^T \frac{dv(t)}{dt} v(t) dt$$

Which can also be written as follows:

$$\Delta E_{kin} = m \int_0^T \frac{d}{dt} \left[\frac{1}{2} v(t)^2 \right] dt$$

(If you're so inclined, to convince yourself of this you can apply the chain rule for differentiation. Those of you who have cast out the spectre of calculus some time ago will just have to take me at my word, I'm afraid.)

And, finally, this becomes:

$$\Delta E_{kin} = \frac{1}{2} m(v(T)^2 - v(0)^2)$$

But the object was dropped from rest, so its starting velocity (the velocity it had at $t = 0$) was 0. And $v(T)$ is the velocity the object has at the end of its journey, which we can just call v, the final velocity. And so our final formula becomes:

$$\Delta E_{kin} = \frac{1}{2} mv^2$$

Okay, so that's the maths—we see that the v^2 falls naturally out of the calculus. But what's the physics behind why we need to square the velocity? A quick Google search shows that this is a fairly popular question, and thus there are quite a few answers, including those in this thread: physics.stackexchange.com/questions/535/why-does-kinetic-energy-increase-quadratically-not-linearly-with-speed.

A nicely intuitive explanation for why kinetic energy doesn't increase linearly with velocity (i.e., why we can't have a formula for the change in kinetic energy that would be something like, say, $\Delta E_{kin} = $ ½ mv) is given by Mike Dunlavey in that Physics StackExchange thread. Dunlavey suggests reframing the question as, "Why does velocity only increase as the square root of kinetic energy, rather than linearly?"

Drop that CD from a height of 1 meter and let's say it has a velocity v when it hits the ground. Now repeat the experiment, but this time drop the CD from a height of 2 meters. As Dunlavey pithily points out, the CD won't hit the ground with a velocity $2v$ because it will cover that second meter in less time than the first.

Where Does the Energy Go in Destructive Interference?

This is another perennial question. The argument is as follows: Energy is conserved, right? So when two waves destructively interfere and "cancel each other out," . . . where does the energy go?

The standard response to this question is to point out, quite correctly, that the energy is redistributed and that the entire wave system needs to be considered rather than focusing on just one point in time and/or space. The total energy is conserved under those conditions.

In a neat paper published in 2014,[11] however, Drosd, Minkin, and Shapovalov analyze a particularly interesting case where the law of energy conservation at first appears to be violated. Imagine we have a pair of sources of fully coherent waves that are in phase with each other (the peaks and troughs are perfectly in step) and have the same amplitude, A. As the authors point out, this gives rise to the following fascinating paradox. The energy of oscillation, E_{one}, at every point of a single wave is given by the equation $E_{one} = \alpha A^2$, where α is a constant (whose value we needn't worry about). But for the two waves together, the amplitude is $2A$ (due to an exceptionally important piece of physics called the principle of superposition). This means that the energy for the two perfectly in-phase waves, E_{two}, is $4\alpha A^2$. In other words, $E_{two} = 4E_{one}$.

So where does that energy come from?

---

[11] R. Drosd, L. Minkin, and AS Shapovalov, *The Physics Teacher* 52 428 (2014).

I enthusiastically recommend that you read the paper in question, if you can access it.[12] But just in case, I'll summarize. The authors first point out that the redistribution of energy explanation doesn't really hold water in this case—if we take the example above and assume the sources of the waves are at the same location, there'll be constructive interference at *all* points. The waves are perfectly in step and originating from the same point in space, after all.

The resolution lies in realizing what happens when two sources get so close together that their separation is lower than the wavelength of the wave. (This is what we call the "near field regime" in nanoscience and it's the basis of a technique that is powerful and infuriating, due to the experimental challenges, in equal measure: scanning near field microscopy.) What happens under those circumstances is that the two sources are coupled together—one affects the other. As Drosd, Minkin, and Shapovalov put it so well:

> *For closely spaced wave sources, it is simply not possible to superimpose the oscillations from two in-phase sources without drawing more energy from the oscillators themselves. In other words, two in-phase sources will "work harder" when near one another, which solves the "paradox" of where the "extra" power comes from for this configuration.*

They use the example of a tuning fork as a pair of oscillators whose separation is smaller than the wavelength of the sound waves. For instance, if we take a tuning fork that produces an A note at a frequency of 440 Hz, the wavelength is ~76 cm. This is of course much, much greater than the separation of the tuning fork tines. For the tuning fork, the interference is destructive rather than constructive, but the

---

[12] Don't get me started on the ludicrousness of the academic publishing industry, where so much work is taxpayer funded and yet unavailable to the public due to exorbitant subscription/article charges. Fortunately, there's been a move to open access publishing of late, which, although not without faults of its own, is at least increasing the availability of academic research.

same arguments hold. The tines of the fork influence each other such that if one tine is covered, the sound from the fork gets *louder* (because destructive interference no longer plays a role).

The Wave Equation

Waves on a string satisfy the wave equation (in one dimension):

$$v^2 \frac{\partial^2 y}{\partial x^2} = \frac{\partial^2 y}{\partial t^2}$$

where v is the speed of the waves on the string, x is the displacement along the string, y is the displacement perpendicular to the string, and t represents time. Note the partial derivative symbols (the "∂"s). We need these because we are considering the motion of the string both in space and time and can therefore differentiate with respect to either: the movement of the string in the y direction is a function both of the position along the string (x) and of time (t).

That equation allows us to model how a wave travels on a guitar string, but, remarkably, it also applies directly to electromagnetic waves. It was part of James Clerk Maxwell's genius that he recognized that light was an electromagnetic wave. Simply change the v for a c and the equation then describes how light propagates: an exceptionally elegant example of just how intimately physics and mathematics are intertwined. And as I mentioned earlier, Maxwell's cross-disciplinary activities didn't stop there—he was also an accomplished poet (who frequently took the opportunity to skewer his peers in verse).

I am not going to derive the wave equation here because that has been done more often over the years than re-releases of Sabbath's (or Floyd's or Zeppelin's or . . .) back catalog. The Hyperphysics website does a very good job of explaining the derivation in a clear and concise way: hyperphysics.phy-astr.gsu.edu/hbase/Waves/waveq.html#c1.

Beyond the Perfect String

If we, like good physicists, assume a perfect string—a string that is infinitesimally thick—then when we use the wave equation above, adding in the boundary conditions for the string (i.e., the fact that it's clamped at the ends), we find that the solutions are precisely the type of harmonics we discussed in Chapter 4. We have solutions that are evenly spaced in frequency so that the second harmonic is twice the frequency of the first, the third harmonic is three times the frequency of the first, and so on.

That type of idealized approach works splendidly in explaining the core concepts because it gets to the key idea of quantized waves: not every frequency is possible and there's a discrete jump from the frequency of one harmonic to that of the next. But this approach also once again places us firmly in the Land of Spherical Cows and Frictionless Floors.[13] If we want to step a little closer to reality, we need to take into account the thickness, stiffness, and stretching of any real guitar string. And when we do that, we modify the wave equation by adding new terms—and that in turn of course leads to different solutions.

For the idealized string, we have a very simple and elegant result (as we saw in Chapter 4):

$$f_n = n f_1$$

In words: the nth harmonic is n times the frequency of the first harmonic (or the fundamental).

When we introduce the effects of bending stiffness, however, the corresponding expression for the frequencies of the harmonics becomes a lot uglier:

$$f_n = \frac{n}{2L} \sqrt{\frac{T}{\rho}} \sqrt{1 + \frac{n^2 \pi^2 E r^4}{4L^2 T}}$$

---

[13] I'm fairly certain that this is the title of a long-lost and unreleased concept album from Peter Gabriel–era Genesis . . .

where L is the length of the string, T is the tension, E is the elastic modulus of the string (a measure of its resistance to being deformed), and r is the radius of the string. Instead of having evenly spaced harmonics, the response of the real string is described as anharmonic—the harmonics no longer have a fixed frequency separation and are characterized instead by that mishmash of variables.

This is the problem with dealing with the real world: it gets very messy, very quickly. But as any fan of the often chaotic live metal experience (where energy and entertainment sometimes outpace studied musicianship) will tell you, there can be beauty in messiness and imperfection, too.

ACKNOWLEDGMENTS

There are so many people I need to thank without whose help, advice, and patience this book would never have seen the light of day. First and foremost, I owe a stadium-sized debt of gratitude to Alexa Stevenson, my editor at BenBella, for taking my tortuous text and kicking it into shape. I learned so very much from Alexa about the art of writing as I watched my long, directionless prose get cut down to size time and again. Alexa also knows that I ascribe to Douglas Adams's philosophy when it comes to deadlines—"I love the whooshing sound as they go flying by"; I am very much the typical academic when it comes to submitting anything on time. She did a remarkable job of not only reining in those academic tendencies but of helping me explain the physics in a style that was much more accessible than what I could ever have achieved without her very many perceptive questions and comments. Laurel Leigh and Scott Calamar also played essential roles in trimming down the progressive rock-esque excesses of my writing. For one thing, they both had the unenviable task of translating my British and Irish colloquialisms for a broader audience.

Despite Alexa's best efforts, I still missed deadlines. (Those academic habits die hard.) And this book would never have made it out of the copyediting stage if it weren't for Lori Ledford's invaluable help in chasing down permissions for the song lyrics and images used throughout the preceding chapters. Thank you, Lori, for your organizational skills that far exceed mine and truly go to 11 (and beyond)! Thanks also to the musicians and scientists who provided those permissions, particularly at very short notice.

Another person who had to continuously deal with my lack of time management was Pete McPartlan, the incredibly talented illustrator I had the immense pleasure of working with for this book. Pete had to regularly handle eleventh-hour requests—often as late as two minutes to midnight—for particular figures (or adaptations to figures). And every time he delivered the goods. In spades. Thanks, Pete!

A number of friends and colleagues read drafts of early versions of the book. Their feedback was extremely helpful in fine-tuning the explanations (and in pointing out just where I'd screwed up a description of the physics). So, a very big thank-you, in alphabetical order, to Alex Allen, Hannah Coleman, Dave Fowler, Sam Jarvis, Filipe Junqueira, Lori Ledford, Chris Morley, Mick Phillips, Philipp Rahe, Michael Rowlands, Jesse Silverberg, Adam Sweetman, and James Theobald.

I also owe Niamh, Saoirse, and Fiachra a big thank-you for putting up with my mantra of "Just let me finish this paragraph/page/chapter" over the last year, as I tapped my laptop keyboard along with the strains of *Operation: Mindcrime*, *Among The Living*, *Pyromania*, *Permanent Waves,* or a variety of other classic albums. (I'll make metal fans of you three yet . . .)

The origins of this book lie in a blog post I wrote for the (now defunct) Institute of Physics blogging project *physicsfocus* back in 2013, with a title almost, but not entirely, the same as that of this book: "When the Uncertainty Principle Goes Up to 11." I had forgone blogging (using the traditional "Who'd have time for *that*?" excuse[1]) until I was contacted by Kelly Oakes, who was the driving force behind

---

[1] In hindsight, and following a few years of fairly intensive social media usage, there's a lot to be said for this particular excuse. I reckon that I spent the best part of half a million words "engaging" online in comments threads over the years. What an utter waste of time and energy. There's more on the pitfalls of social media engagement for academics here: http://blogs.lse.ac.uk/impactofsocialsciences/2017/05/02/rules-of-engagement-seven-lessons-from-communicating-above-and-below-the-line/.

the *physicsfocus* project (and is now Science Editor for BuzzFeed UK). Without Kelly's encouragement (alongside that of Louise Mayor, Features Editor for *Physics World* at the time), I'd never have started blogging, the "Uncertainty Principle to 11" post wouldn't have existed, and, in turn, this book would never have appeared. Brady Haran must also take some of the blame. I have learned so much from Brady over the last decade (gulp) about pitching science to audiences well outside my traditional academic comfort zone. At times, I swear I could hear his dulcet Australian tones behind me when I was needlessly making hard work of the writing: "Why should I care about this, Phil? Why should *anyone* care about this?"

Special thanks goes out to Stan Wakefield, the literary agent who introduced me to BenBella, and, of course, to Glenn Yeffeth, who patiently listened to me overenthusiastically and rather incoherently pitch the metal-quantum physics "nexus" and still had enough confidence in the concept to commission "When The Uncertainty Principle . . ." as a book. Thank you, Glenn, and thanks to all at BenBella (alongside Alexa, Laurel, and Scott) who have contributed in many ways to the development and marketing of this book: Adrienne Lang, Jennifer Canzoneri, Alicia Kania, Monica Lowry, and Sarah Avinger. It's been a pleasure working with you all.

Finally, it would be remiss of me not to acknowledge the inspiration of the very many metal (and rock) bands whose music played (very loudly) in the background as I wrote this book. I'll leave the final words to Neil Peart of Rush, whose music has featured as the backdrop to so much of my life. Peart's lyrics in "The Spirit of Radio" are a great example of what "When The Uncertainty Principle Goes to 11" is all about: bridging that frustrating cultural gap that still too often exists between the arts and science. "*All this machinery making modern music can still be open-hearted. Not so coldly charted. It's really just a question of your honesty . . .*"

INDEX

ABOUT THE AUTHOR

Philip Moriarty is a physicist, a heavy metal fan, and a keen air-drummer.

He is a professor at the University of Nottingham, where his research focuses on prodding, pushing, and poking single atoms and molecules; in this nanoscopic world, quantum physics is all. He's also a writer, a TEDx speaker, and a regular contributor to the Sixty Symbols YouTube project. His work has appeared in *The Independent*, *The Guardian*, *Times Higher Education*, *BBC Radio 4*, *Physics World*, and *The Economist*, amongst others.

Moriarty has taught physics for almost twenty years and has always been struck by the number of students in his classes who profess a love of metal music, and by the deep connections between heavy metal and quantum mechanics. His three children—Niamh, Saoirse, and Fiachra—have patiently endured his off-key attempts to sing along with Rush classics for many years. Unlike his infamous namesake, Moriarty has never been particularly enamored of the binomial theorem.